Green Chemistry and Sustainable Technology

Aims and Scope

The series Green Chemistry and Sustainable Technology aims to present cutting-edge research and important advances in green chemistry, green chemical engineering and sustainable industrial technology. The scope of coverage includes (but is not limited to):

- Environmentally benign chemical synthesis and processes (green catalysis, green solvents and reagents, atom-economy synthetic methods etc.)
- Green chemicals and energy produced from renewable resources (biomass, carbon dioxide etc.)
- Novel materials and technologies for energy production and storage (bio-fuels and bioenergies, hydrogen, fuel cells, solar cells, lithium-ion batteries etc.)
- Green chemical engineering processes (process integration, materials diversity, energy saving, waste minimization, efficient separation processes etc.)
- Green technologies for environmental sustainability (carbon dioxide capture, waste and harmful chemicals treatment, pollution prevention, environmental redemption etc.)

The series Green Chemistry and Sustainable Technology is intended to provide an accessible reference resource for postgraduate students, academic researchers and industrial professionals who are interested in green chemistry and technologies for sustainable development.

More information about this series at https://link.springer.com/bookseries/11661

Neeraj K. Aggarwal · Naveen Kumar ·
Mahak Mittal

Bioethanol Production

Past and Present

Springer

Neeraj K. Aggarwal
Department of Microbiology
Kurukshetra University
Kurukshetra, Haryana, India

Naveen Kumar
Department of Microbiology
Kurukshetra University
Kurukshetra, Haryana, India

Mahak Mittal
Department of Microbiology
Kurukshetra University
Kurukshetra, Haryana, India

ISSN 2196-6982 ISSN 2196-6990 (electronic)
Green Chemistry and Sustainable Technology
ISBN 978-3-031-05090-9 ISBN 978-3-031-05091-6 (eBook)
https://doi.org/10.1007/978-3-031-05091-6

This Springer imprint is published by the registered company Springer Nature Switzerland AG
The registered company address is: Gewerbestrasse 11, 6330 Cham, Switzerland

Dedicated to Scientific Community

Green Chemistry and Sustainable Technology

Aims and Scope

The series Green Chemistry and Sustainable Technology aims to present cutting-edge research and important advances in green chemistry, green chemical engineering and sustainable industrial technology. The scope of coverage includes (but is not limited to):

- Environmentally benign chemical synthesis and processes (green catalysis, green solvents and reagents, atom-economy synthetic methods etc.)
- Green chemicals and energy produced from renewable resources (biomass, carbon dioxide etc.)
- Novel materials and technologies for energy production and storage (bio-fuels and bioenergies, hydrogen, fuel cells, solar cells, lithium-ion batteries etc.)
- Green chemical engineering processes (process integration, materials diversity, energy saving, waste minimization, efficient separation processes etc.)
- Green technologies for environmental sustainability (carbon dioxide capture, waste and harmful chemicals treatment, pollution prevention, environmental redemption etc.)

The series Green Chemistry and Sustainable Technology is intended to provide an accessible reference resource for postgraduate students, academic researchers and industrial professionals who are interested in green chemistry and technologies for sustainable development.

More information about this series at https://link.springer.com/bookseries/11661

Neeraj K. Aggarwal · Naveen Kumar ·
Mahak Mittal

Bioethanol Production

Past and Present

Springer

Neeraj K. Aggarwal
Department of Microbiology
Kurukshetra University
Kurukshetra, Haryana, India

Naveen Kumar
Department of Microbiology
Kurukshetra University
Kurukshetra, Haryana, India

Mahak Mittal
Department of Microbiology
Kurukshetra University
Kurukshetra, Haryana, India

ISSN 2196-6982 ISSN 2196-6990 (electronic)
Green Chemistry and Sustainable Technology
ISBN 978-3-031-05090-9 ISBN 978-3-031-05091-6 (eBook)
https://doi.org/10.1007/978-3-031-05091-6

This Springer imprint is published by the registered company Springer Nature Switzerland AG
The registered company address is: Gewerbestrasse 11, 6330 Cham, Switzerland

Dedicated to Scientific Community

Preface

Emerging issues of the energy crisis, as a consequence of industrial advancement and urbanization, global warming and climate changes, diminishing oil resources and its increasing prices are big concerns worldwide. The world's present energy supplies rely on fossil fuel sources. But extensive and rapid exploitation of existing fossil fuel reserves by increasing world population will lead to their exhaustion in the approaching decade bioethanol is an alcohol made through the fermentation of plant sugars from agricultural crops and biomass resources. Bioethanol in recent years has emerged as one of the alternative liquid fuels and has generated immense activities of research on the production of ethanol and its environmental impact. Bioethanol, among prevalent biofuels, is the major contributor to the transport sector. This can be either used directly replacing existing transport fuels completely or blended with petrol and diesel. Government regulations in several countries have made ethanol blends mandatory in transport vehicles in different ratios, i.e., E10 to E85. Blending of bioethanol with petroleum-based transport fuels will not only impede petroleum import in nations but also mitigate GHGs emissions.

The book provides an updated and detailed overview of past and present scenarios in bioethanol production. This book comprehensively covers the global scenario of ethanol production from both food and non-food crops and other sources. It looks at the historical perspectives, chemistry, sources, and production of ethanol and discusses biotechnology breakthroughs and promising developments through nanotechnology and biorefinery concepts. The book also provides important insights into resource management and the environmental and economic impact In addition, it presents information about ethanol in transport sector and also highlights the challenges and future of ethanol.

Kurukshetra, India

Neeraj K. Aggarwal
Naveen Kumar
Mahak Mittal

Contents

About the Authors

Dr. Neeraj K. Aggarwal is presently working as Associate Professor and Chairman, Department of Microbiology, Kurukshetra University Kurukshetra, India. He is having more than 15 years of teaching experience of PG classes. He has guided 13 Ph.D. research students and presently seven students are working under his guidance. He has obtained his M.Sc. and Ph.D. in Microbiology from CCS Haryana Agricultural University Hisar. He is gold medalist and has been awarded prestigious S. R. Vyas gold medal of CCS Haryana Agricultural University for being best microbiological research worker. He has published more than 120 research papers, nine review articles and 14 book chapters in various journals of national and international repute. In accordance to Google Scholar, his work has received total citations of over 2000 with an h-index of 22 and i10 index of 39 to date He has one ongoing research project on bioethanol production from *Parthenium hysterophorous* and completed three research projects awarded by different funding agencies. He has been awarded INSA Summer Research Fellowship. He also authored *Water Hyacinth: A Potential Lignocellulosic Biomass for Bioethanol* (with Springer), a popular text book *Introduction to Biotechnology* for B.Tech. Students, two edited book *Microbiology and Biotechnology for a Sustainable Environment* and *Aptamers* (with Springer).He is member of various academic and Professional bodies. Broadly, Dr. Aggarwal has focused his research areas around microbial biotechnology for production of different metabolites, molecular genetics and biological control agents. His lab is engaged in utilization of different industrial wastes for some value added products like and bioplastics. He and his team is also involved in exploitation of different agricultural residues of lignocellulosic nature for bioethanol production. Different medicinal plants were also screened for their antimicrobial potential and other health benefits.

Naveen Kumar is member of biofuel group of fermentation laboratory, Department of Microbiology Kurukshetra University Kurukshetra. He has obtained his master degree in Microbial Biotechnology from MD University Rohtak and is currently pursuing his Ph.D. on bioethanol production from *Parthenium hysterophorus.*

Mahak Mittal is a Ph.D. research scholar in the Department of Microbiology, Kurukshetra University, Kurukshetra. She has recieved Gold Medal in her Masters in Microbiology from Kurukshetra University, Kurukshetra. She has been awarded prestigious RUSA fellowship of the university for her Ph.D. research work. Currently, she is working on biovalorization of domestic waste for some value added products.

Chapter 1
Bioethanol: An Overview of Current Status and Future Direction

Abstract As a biofuel, bioethanol is one of the most intriguing because of its beneficial environmental impact. It is currently made primarily from sugar and starch-containing raw ingredients. Most biofuels may be used either as a stand-alone fuel or in a blend with gasoline. Ethanol is by far the most common type. Raw materials for ethanol manufacturing include sugars, starches, and lignocellulosic materials, as well as algae biomass. This chapter addresses overview of several aspects of bioethanol production, such as worldwide trends, diverse raw materials, the status and recent advancements of applicable technologies, biorefinery aspects, as well as the benefits and drawbacks of fuel ethanol. Currently, global bioethanol output is at an all-time high. Bioethanol looks to have a bright future because there is a strong need for sustainable energy sources to decrease reliance on foreign oil.

Keywords Global trends · Bioethanol · Bioethanol production

1.1 Introduction

Countries all over the world have well-thought-out and directed state policies aimed at increasing and cost-effective biomass utilization to satisfy future energy demands and meet the Kyoto Protocol's CO_2 reduction targets, as well as to reduce reliance on fossil fuel supply [1]. Biomass-based fuel bioethanol production appears to be a promising substitute for conventional fossil fuel sources of energy. Bioethanol is primarily produced by yeast fermentation from agricultural residues as an alternate source of energy. The most recognizable renewable energy source is lignocellulosic material [2]. Bioethanol is not a new energy source; it was widely employed in Europe and the United States in the early 1900s, but it was overlooked due to its higher manufacturing costs than gasoline. Due to the 1970s oil crisis, production of bioethanol continued [3]. Bioethanol, when compared to gasoline, can reduce CO_2 emissions by up to 80%, resulting in a cleaner environment in the future [4]. Ethanol can be used as a sole fuel in specialized engines or as an octane enhancer in mixtures. Sugars, starches, and cellulosic materials are currently used to produce bioethanol. The world's greatest biomass source is cellulose materials generated from forest and agricultural residues, municipal solid wastes, or energy crops [5].

N. K. Aggarwal et al., *Bioethanol Production*, Green Chemistry and Sustainable Technology, https://doi.org/10.1007/978-3-031-05091-6_1

1.2 Global Trends

The United States is the world's largest producer of fuel ethanol. The United States is expected to produce 13.8 billion gallons of ethanol in 2020. The United States is also the world's biggest producer of ethanol. Up to 10% of ethanol, grain alcohol, can be combined with gasoline and used in regular automobiles [6]. For the first time ever, India Glycols Ltd. in Kashipur, Uttarakhand, has built a second-generation ethanol demonstration unit that uses a wide variety of feedstocks, including rice straw, sugarcane bagasse, and cotton stalk. The construction of a prototype plant with a feeding capacity of 50 kg per day on the campus of the National Institute for Interdisciplinary Science and Technology (NIIST) is part of an ongoing effort to produce second-generation ethanol [7] (Fig. 1.1).

With a 17% increase from 2009, the global production of biofuels in 2010 was 28 billion US gallons, accounting for 2.7% of the world's highway transportation fuel supply. Bioethanol and biodiesel accounted for the majority of this contribution [8]. China was Asia's top ethanol producer in 2012, producing 2.1 billion liters of gasoline-ethanol. Ethanol has been developed in China for use as a car fuel for some time. China intends to promote ethanol-based gasoline and create a new market for its surplus grain in order to reduce oil consumption. Corn, cassava, and rice are the most common sources of grain-based bioethanol in China, accounting for around 80% of total bioethanol production [9]. Bioethanol and biodiesel were approved

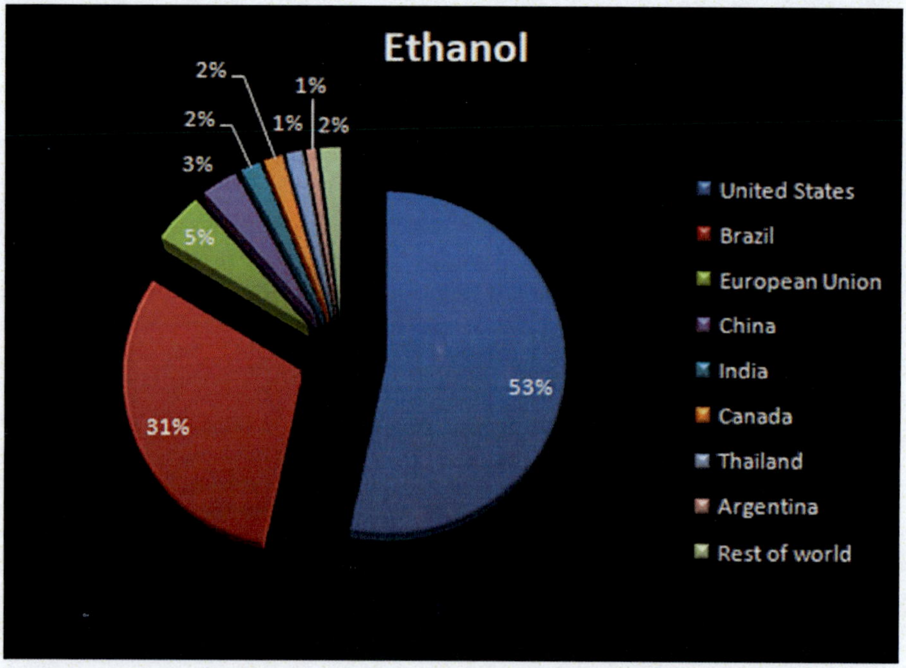

Fig. 1.1 Global ethanol production by country in 2020 (millions of gallons) [6]

as biofuels by the Mexican Congress in 2008 under the Bioenergetics Act [10]. It took 7.56% of Mexico's total energy production to come from renewable sources in 2014. Biomass feedstocks made up only 4.12% of this renewable energy [10]. They obtained ethanol yields of 58 and 48 (gallon/metric ton) for blue agave and sugarcane bagasse, respectively, despite poor efficiency in all steps [11]. The United States exported 6.5 billion liters of ethanol in 2018, increasing its share of global ethanol exports to 61%. In the United States, starch-based crops are processed in dry or wet mills to produce the bulk of ethanol. The US will have 208 ethanol-producing units in 2020, with a total installed capacity of 17.44 billion gallons per year. In 2020, the overall production was 13.8 billion gallons, down 12.7% from the previous year's record of 15.8 billion gallons [12, 13]. Canada was classified as the world's 6th largest bioethanol producer in 2020. Canada's ethanol output accounted for 1.6% of worldwide production. Corn and wheat were the two most important feedstocks for the production of bioethanol, contributing 1534.3 million L and 360.7 million L, respectively [13]. Brazil's ethanol output will account for 26.7% of worldwide ethanol production in 2020 [14]. In 2020, Argentina ranked eighth in the world, producing 1.0% of total global ethanol output [14]. Colombia was ranked 13th globally in terms of ethanol production in 2020, contributing 0.44% to total worldwide ethanol production [15]. In 2020, the EU's total ethanol output accounted for 4.8% of the world's production [14]. In terms of biofuel production, Vietnam has consumed a significant amount of gasoline in recent years as its economy has grown. According to the United States Department of Agriculture (USDA), Vietnam's gasoline usage has risen by 4–5% every year in recent years. In addition to its production, it is estimated that Vietnam spent $2.5 billion on crude oil and petroleum products, including bioethanol, in the first quarter of 2020 [16]. India came in sixth place among the world's top ethanol producers. India's ethanol production is projected to reach 3.17 billion liters in 2021, up 7% from 2020 due to excess sugarcane production, and the country's average ethanol blending rate in gasoline is expected to reach 7.5 percent in 2021, thanks to increased government initiatives to divert more raw materials toward ethanol [17].

1.3 Bioethanol Production

1.3.1 1G Bioethanol

Food crops high in starch and sugar are used in the 1G of bioethanol production. This is a well-known technique that is commonly employed for the commercial generation of bioethanol. The process of producing bioethanol from sugar crops is extremely straightforward and does not necessitate any lengthy steps [18]. Sugars are extracted from the plants through milling, and the sugars are then fermented to produce bioethanol (Fig. 1.2). The conversion of disaccharides (such as sucrose) into monomers requires enzyme invertase [19]. On the other hand, starch must be

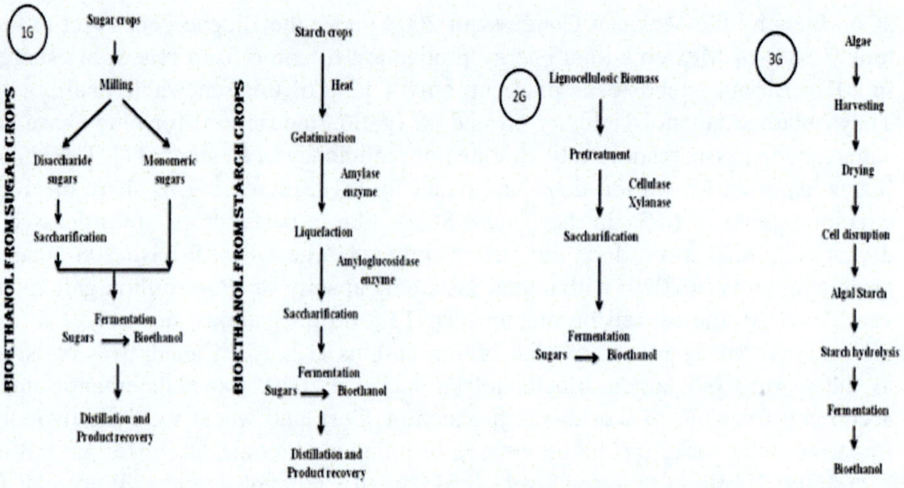

Fig. 1.2 Outline of the 1G, 2G, and 3G bioethanol production processes

gelatinized or liquefied and hydrolyzed to produce bioethanol (Fig. 1.2). Gelatinization is a method that involves heating food with water to cause starch granules to expand, allowing the starch to be extracted. The starch-rich slurry is liquefied using amylase, an enzyme that catalyzes the digestion of long amylose polymers of starch into short oligosaccharides. Following that, treatment with glucoamylase and amyloglucosidase transforms oligomers into monomeric sugars during saccharification. Finally, an ethanologenic microbial strain is used to convert sugar monomers to bioethanol. Current bioethanol supplies are made using 1G bioethanol production technology that uses sugar or starch crops. Brazil and the United States, the world's major bioethanol producers, manufacture bioethanol from sugarcane and corn, respectively, accounting for over 70% of global bioethanol output [1]. As a result, large agricultural countries such as the United States, Brazil, and China are currently the primary producers of ethanol, including alcohol fuels. In the United States, corn grain starch has been used to manufacture 95% of bioethanol [20].

1.3.2 2G Bioethanol

Lignocellulosic material has been recommended as a sustainable feedstock for bioethanol production in the search for renewable and non-food materials [1, 21]. Cellulosic biomass has been reported to have a better potential for bioethanol generation than starch and sugar crops [21, 22]. Cellulosic bioethanol, sometimes called 2G bioethanol, is produced from lignocellulosic materials. The most abundant renewable resource on the planet is lignocellulosic cellulose [23]. It is the most important structural component of all crops [24] and accounts for over half of the world's biomass [25]. By 2022, the United States' Energy Independence and Security Act of 2007 aimed to produce 36 billion gallons of renewable fuel, with 16 billion gallons coming from cellulosic bioethanol [26]. Converting lignocelluloses to bioethanol is substantially more complicated than sugar or starch conversion. Pretreatment, saccharification, fermentation, and product recovery are the four processes in the biological process for turning lignocellulosic material into bioethanol fuel [27] (Fig. 1.2). The pretreatment of the material is the most difficult processing hurdle in the manufacturing of cellulosic ethanol. Pretreatment is a necessary step for exposing the carbohydrates hidden in the lignocellulose core, which is made up of cellulose and hemicellulose encased in a matrix of a very recalcitrant substance called lignin [28]. Pretreatment involves removing tough lignin completely or partially using a variety of ways. As a result, the pretreatment phase adds to the ethanol production process's complexity and cost. The carbohydrate polymers of cellulose or hemicellulose are hydrolyzed to their respective monomers, i.e., hexose and pentose sugars, which are then transformed to bioethanol in the subsequent fermentation stage.

1.3.3 3G Bioethanol

Algal biomass has recently piqued the interest of researchers and industry leaders as a potential alternative feedstock for biofuels such as bioethanol, biodiesel, biogas, photobiologically produced biohydrogen, bio-oil, and syngas [29]. Algal biomass is used to make 3G bioethanol. Algae are divided into two categories: microalgae and macroalgae. Because of their high lipid content, microalgae have gotten a lot of attention for biodiesel manufacturing. Macroalgae, on the other hand, are utilized to produce bioethanol. Figure 1.2 depicts the process of producing ethanol from algae. According to Schenk et al. [30], a hectare of algal biomass may produce 365 tons of dry weight per year. Furthermore, due to its low hemicellulose and lignin content, algal biomass appears to be a viable source of cellulosic ethanol. As a result, algae offer a huge potential for commercial bioethanol production [31].

Bioethanol is usually made from sugarcane juice or molasses, both of which come from the sugarcane business. Fresh sugarcane juice accounts for about 79% of the ethanol produced in Brazil, with cane molasses accounting for the leftover percentage. In India, sugarcane molasses is the primary substrate for the production of bioethanol; cane juice is not currently utilized [30]. Bioethanol production could be increased with the use of bagasse, the solid waste left over from juice extraction. There is an issue about how long and how much food crops can be used as bioethanol building materials. This comes from a rise in human populations and the use of food crops in 1G bioethanol production. Furthermore, this issue has the potential to have a significant impact on the pricing of present bioethanol feedstocks. As a result, non-food materials will take precedence and will continue to be the only sustainable source of renewable bioethanol feedstock [32]. Because of their higher levels of fermentable sugars and total carbohydrates, 1G and 2G feedstocks like sugarcane and cereals are the most commonly used feedstocks in the industrial-scale production of bioethanol [33]. Growing feedstock crops requires a large amount of arable land and causes plants to grow more slowly, which is a significant obstacle to the advancement of bioethanol production. Microalgae is a better choice than traditional materials because it has a lot of advantages over them, like a faster rate of production and growth, less space, less time to harvest, and only a simple nutrient need for growth [34]. In general, microalgae strains that grow quickly and produce a lot of biomass are desirable biofuel possibilities. *Chlorella vulgaris* and *Chlamydomonas reinhardtii* can produce up to 3.28 g L^{-1} and 0.86 g L^{-1} of biomass per day, which makes them attractive candidates for biofuel production [35, 36]. Because algae have a higher photosynthetic efficiency than land plants, their biomass production is 5–10 times higher [37]. Lignin, a physical barrier to enzyme hydrolysis that cannot be removed through pretreatment, is absent in algae. This property of algae is beneficial in the production of ethanol during pretreatment and enzymatic hydrolysis [38]. However, the commercialization of microalgae-derived biofuels is still hindered by slow growth rates and low biomass production. It has been used in many countries as a fraction and blends of petroleum products [39].

1.4 Industrial Scenario of Bioethanol-Based Sanitizers

The market for ethanol-based hand sanitizers has grown at an exponential rate over the past few years. The global coronavirus outbreak of 2019 has resulted in a 1400% increase in demand for hand sanitizers from December 2019 to February 2020. According to a new report from Fior Markets [40], hand sanitizer sales are expected to increase from 1.2 billion USD in 2019 to 2.14 billion USD in 2027 [40]. The Red River Biorefinery (RRB) in North Dakota, which produces ethanol from 500,000 tons of agricultural byproducts per year, has recently begun generating pharmaceutical grade (USP) ethanol for cleaning agents and hand sanitizers [41].

1.5 Use of Bioethanol as Transport Fuel and Challenges

Ethanol has several advantages. It biodegrades quickly in the environment and emits far fewer pollutants than petroleum-based fuels in combustion engines [42, 43]. Significantly, ethanol can be utilized in two ways: either as a fuel additive (E10) or as a virtually stand-alone fuel (E85). E10 is also known as gasohol. It is a fuel blend of 10% anhydrous ethanol and 90% gasoline that may be used in the ICE of most modern vehicles and light trucks without requiring any engine or fuel system changes. In the United States and Europe, E85 is the most ethanol-rich fuel mixture, and it is the standard fuel for flexible-fuel vehicles. Currently, E85, a blend of 85% ethanol and 15% gasoline, is widely used in flexible-fuel vehicles (FFVs) and variable fuel vehicles (VFVs), and many manufacturers (including Ford, General Motors, and Chrysler Corporation) offer vehicles that can run on 100% gasoline, E85, or any combination of the two (estimated to be over 7 million FFVs on the roads by

2009). The octane number of ethyl alcohol is 112, while E85 has an octane level of around 105. Because of the high-octane rating, ethanol fuel can be used in even high-performance engines. Ethanol is an excellent racing fuel because it burns at a lower temperature than gasoline, meaning it needs less radiator cooling power. Ethanol is produced from crops that absorb CO_2 and release oxygen. This helps to keep the CO_2 levels in the atmosphere in check. Increasing the use of renewable fuels like bioethanol will help to minimize pollution and global warming caused by gasoline combustion. Under current conditions, the use of ethanol fuel blends like E85 can reduce total greenhouse gas emissions by as much as 30–36% while also reducing fossil energy use by 42–48% [44]. Ethanol-blended fuel (E10) reduces greenhouse gas emissions by 2.4–2.9% while decreasing fossil energy consumption by 3.3–3.9%.

The stability of mixes is one of the most difficult difficulties with ethanol fuel blends. Because of their water-absorbing and corrosive properties, ethanol fuel blends have a substantially shorter shelf life. Without the use of a stabilizer, it cannot be stored for more than two to three months [45, 46]. Ethanol has a solvent action, which means it will remove gums and other deposits in fuel systems that have been running on mineral gasoline for a long time. In extreme cases, this can block fuel filters and cause the engine to run poorly. Because gasoline is normally taken from the bottom of the tank, the water/ethanol layer gets sucked into the engine first, preventing it from running [47, 48]. Ethanol production on a small scale necessitates a large investment in both equipment and labor. Processing the alcohol to the requisite 200 proof for mixing with gasoline in unmodified engines adds to the expense. Ethanol should be used in upgraded engines to be the most effective. Mixing ethanol and gasoline makes less sulfur dioxide and CO2, but it also makes more nitrogen oxide and volatile organic compounds (VOCs), which make the ozone less dense [49].

1.6 Biorefinery with Bioethanol Production: Beyond Fuel

Biorefinery is a broad term that encompasses diverse sectors, like, chemical, transport, energy, and agriculture, and therefore it is challenging to make a single definition. For example, according to the International Energy Agency (IEA), biorefining is the sustainable conversion of biomass into broad range of bio-based products and biofuels, as an efficient scheme to employ available feedstock for the synergistic production of energy. According to another definition, biorefinery is defined as "a facility for the synergetic conversion of biomass into multiple commercial bio-based products (food and feed ingredients, chemicals, materials, minerals, CO_2) and bioenergy (fuels, power, heat)" [49]. Studying the costs of the various components of the entire process can be used to estimate the cost of any established process. These components include the cost of the biomass, apparatus, chemicals, and energy consumption during the process in the case of biomass pretreatment. Different scientific groups have devised different pretreatment procedures at the laboratory scale and used mass balance analysis to estimate the overall yield. Flores-Gómez et al. [50]

analyzed the overall process of converting agave wastes to ethanol using mass balance analysis. They found that 1000 kg of A. tequilana leaf yields 198.4 kg ethanol, as well as 21.6 kg residual xylose, 14.2 kg glucose, 22.0 kg xylo-oligomers, and 29.5 kg galacto-oligomers. The whole method entails pretreatment with AFEXTM, followed by 72 h of simultaneous hydrolysis and fermentation with a cocktail of commercial enzymes CTec3 (6.2 kg protein), Htec3 (1.5 kg protein), and yeast. Kazi et al. [51] conducted a techno-economic analysis of biochemical ethanol production techniques using several technologies. They discussed the four main pretreatment procedures used to convert maize stover to gasoline (dilute acid, two-stage dilute acid, AFEX, and hot water pretreatment). The dilute acid approach produced the maximum ethanol production of 289 L/t at the lowest fuel cost of 1.36 $/L of gasoline-equivalent. They also advised that the dilute acid pretreatment's operating cost be expressed as a percentage of the entire operating cost. In terms of feedstock, variable operational cost, fixed operational cost, capital depreciation, and average yearly tax, they showed 35.3%, 37.8%, 6.0%, 10.0%, and 10.9%, respectively. The use of waste substrate to generate power makes the overall process more cost-effective and allows for the collection of a 7.1% electricity credit. Based on the findings, we can conclude that the cost estimation of the total process at the laboratory size is restricted to the raw material used, the procedure employed, and the energy required. As a result, a low-cost pretreatment approach is a key quality in lignocellulosic feedstock-based biorefineries that controls the overall process. During the early days of fossil fuel use, humans used the abundant resources without considering the social consequences, such as the environment and health. Due to increased environmental degradation and global warming over the last four decades, the public and scientific communities have become more conscious of the harmful effects of excessive fossil fuel use. As a result, comparable errors cannot be committed in the development of a lignocellulosic biomass-based biorefinery. As a result, in addition to a techno-economic evaluation, the socioeconomic impact of the process must be investigated. The possibility for local farmers' communities, employment opportunities, environmental impact due to pollutant release (if any), health impact on workers and people living nearby, economic growth, and energy security are some of the socioeconomic indicators. The socioeconomic evaluation aids in the analysis of the social impact of a biorefinery's construction in a region [52].

1.6.1 Value-Added Chemicals

Rass-Hansen et al. [53] have examined bioethanol as a substrate for the industrial applications, stating that bioethanol may be used to produce acetaldehyde, ethylene, butadiene, and acetic acid, among other things. Torula yeast, made by Amoco, is a well-known example of microbial protein products created from ethanol [54]. Other food and feed additives, including amino acids and organic acids, were also investigated. The use of ethanol as a carbon source for the synthesis of L-glutamic acid has been investigated [55]. The ability to accumulate L-glutamic acid seems to be widespread among microorganisms such as *Brevibacterium* sp., *Corynebacterium* sp., and *Pseudomonas*, *Alcaligenes*, and *Bacillus* strains. The greatest concentration of L-glutamic acid obtained was 53.1 g/l, yielding 0.6 g/g ethanol. The prospect of ethanol as a source of carbon for the manufacture of L-lysine was examined by the Japanese firm Kyowa Hakko Kogyo. L-lysine concentrations of 2.1 to 8.4 g/l were produced using *C. glutamicum*, *Brevibacterium ammoniagenes*, *Bacillus megaterium*, *Arthrobacter paraffineus*, and *Nocardia strains*, with a fairly low maximum yield of 0.02 g/g ethanol [56]. Another patent from Mitsubishi Petrochemical Co Ltd claims that *Acinetobacter calcoaceticus* strain YK-1011 was able to produce a tenfold greater yield of 0.2 g L-lysine per g substrate during ethanol fermentation [57]. Vinegar manufacture is a prime example of a two-stage process, with bakers' yeast initially converting sugar to alcohol, followed by a second fermentation process in which the alcohol is transformed to acetic acid by, for example, *Acetobacter* species [58].

1.6.2 Nanomaterials

Nanomaterials are structures with a size varying from 1 to 100 nm (nm). They can be made in a variety of shapes and sizes [59]. Catalysts can be made from a variety of materials, including cobalt and nickel, as well as metal oxides [60, 61]. Magnetic nanoparticles (MNPs) have considerable promise for biofuel and bioenergy applications, such as the immobilization of enzymes like cellulases and hemicellulases on magnetic supports for the production of sugars and bioethanol from lignocellulosic biomass. It is possible to magnetically retrieve and recycle these immobilized enzymes for reuse [62].. In specific quantities, supplemental nanoparticles can boost fermentative ethanol production and change the microorganism's metabolic route to promote ethanol production [63, 64]. According to El-Kemary et al. [65], the hydrophobic component of $C_6H_{12}O_6$ was adsorbed onto the surface layers of nanoparticles by Vander Waals interactions, while the hydrophilic component of the OH is oriented toward the aqueous phase. As this study found, the reaction between the nanoparticles and the sugar made it easier for *S. cerevisiae* BY4743 to take in glucose, which led to more bioethanol production. Fe_3O_4 nanoparticles significantly increased ethanol production, with a highest bioethanol yield of 0.26 g/g, hexose sugar consumption of 99.95%, ethanol productivity of 0.22 g/L/h, and fermentation efficiency of 51% at 0.01 wt% [66].

1.6.3 Weed Biomass

Weeds can be profitable and valuable when utilized as a potential resource. Next-generation biofuel feedstocks can be made from weeds because of their rapid reproduction, abundance of cellulose, and low lignin content. Kuhad et al. used SHF of

L. camara to produce ethanol, with a highest bioethanol output of 148 g/1000 g of pretreated *L. camara* biomass [67]. After being pretreated with 4% H_2SO_4, water hyacinth biomass was saccharified with commercial cellulase enzyme (30 FPU/g biomass) and fermented with *S. cerevisiae*, resulting in an ethanol concentration of 0.292% (w/v) and a process efficiency of 59.3% [68]. After fermenting hydrolysate from NaOH pretreated *Pennisetum polystachion* with *S. cerevisiae* (TISTR 5596), Prasertwasu [69] found a good bioethanol yield after 24 h. Another study utilizing a traditional strain, *S. cerevisiae* NBRC 2346, demonstrated a high yield of 14.9 g/1 ethanol from water lettuce [70].

1.6.4 Municipal Solid Waste

Municipal solid waste management is a big problem all over the world. The improper management of municipal solid waste (MSW) can be extremely dangerous. Researchers from all around the world are working on a plethora of methods to reduce waste production and oversee waste management for long-term socioeconomic and environmental benefits. The framework for analyzing municipal solid waste to energy conversion and waste-derived bioeconomy in order to meet sustainable development goals is very important. *S. cerevisiae* and *Z. mobilis* are more economically viable than *E. coli* and *S. pombe* for producing bioethanol from MSW. Based on several aspects such as autoclave pretreatment, reduced enzyme consumption, and increased solid loading, viable ethanol production from MSW can be achieved [71]. A study has optimized pectin, ethanol, and biogas production. From 1 t of orange trash, the highest yield was 244 kg of pectin, 26.5 L of ethanol, and 36 m3 of methane [72]. The integrated MSWM facility is both environmentally and economically sustainable, with secondary product replacements offsetting environmental and financial costs (e.g., recycled plastics, bricks, and metals). In terms of cost, replacing coal gas, natural gas, and electricity with biogas has the potential to save 11.8%, 18.9%, and 8.5% of the overall cost, respectively, as compared to coal replacement [73].

1.6.5 Agricultural Biomass

Agricultural wastes can produce a large amount of renewable energy due to their lignocellulosic properties [74]. Agricultural wastes containing starch, such as cassava peels and potato peels, are often generated in massive quantities every year all over the world. They are a collection of important polysaccharide compounds that can be used to make biofuels and bioenergy. Through the implementation of suitable solutions aimed at boosting product yield and process feasibility, starch-based agricultural wastes can be studied for bioethanol production. Research on the use of agricultural wastes for the production of second-generation bioethanol has so far produced very promising results worldwide. Several studies have been reported on lab-scale and pilot scale for the production of bioethanol from agricultural waste but still, there is a gap between lab-scale production and industrial production. Therefore, to utilize these cheap, abundantly present, and renewable resources for sustainable production of bioethanol, different hurdles have yet to be overcome. These include-easy and cost-effective process of pretreatment, efficient fermentation by employing ideal microorganisms, and economic product recovery. Considering huge abundance of agricultural feedstock, various efforts are being carried out to make second-generation bioethanol production more sustainable and cost-effective.

1.7 Summary

Corn ethanol in the US and sugarcane ethanol in Brazil have both been in commercial use for a long time, so there is not much room for further research to improve the technology. COVID-19 has swept the world, harming the world's manufacturing industries. As a result, global bioethanol output fell in 2020. A rebound is projected when the epidemic is finished and the production industries return to normal. However, production may not be as strong as it was before the global lock-down phase. After restrictions on making hand sanitizer were lifted during the pandemic, the demand for ethanol for making hand sanitizer is still high. The most difficult problem is lowering the cost of producing bioethanol. As a result, the biorefinery idea is required to more completely utilize renewable fuel sources and to produce additional value-added products (e.g., bio-based products from lignin) that reduce ethanol production costs. As a result, ethanol will be less expensive than traditional fossil fuels.

References

1. Sarkar, N., Ghosh, S. K., Bannerjee, S., & Aikat, K. (2012). Bioethanol production from agricultural wastes: An overview. *Renewable Energy, 37*(1), 19–27.
2. Haveren, J. V., Scott, E. L., & Sanders, J. (2008). Bulk chemicals from biomass. *Biofuels, Bioproducts and Biorefining: Innovation for a Sustainable Economy, 2*(1), 41–57.
3. Balat, M. (2009). New biofuel production technologies. *Energy Education Science and Technology Part A, 22*(2), 147–161.
4. Lashinsky, A., & Schwartz, N. D. (2006). How to beat the high cost of gasoline. *Fortune, 153*(2), 74
5. Lin, Y., & Tanaka, S. (2006). Ethanol fermentation from biomass resources: Current state and prospects. *Applied microbiology and Biotechnology, 69*(6), 627–642.
6. Sönnichsen, N. (2021). *Fuel ethanol production worldwide in 2020, by country*. Retrieved from statista: https://www.statista.com/statistics/281606/ethanol-production-in-selected-countries/
7. Sukumaran, R. K., Surender, V. J., Sindhu, R., Binod, P., Janu, K. U., Sajna, K. V., Rajasree, K. P., & Pandey, A. (2010). Lignocellulosic ethanol in India: Prospects, challenges and feedstock availability. *Bioresource Technology, 101*(13), 4826–4833.
8. Biofuels make a comeback despite tough economy. World Watch Institute. ⟨http://www.worldwatch.org/biofuels-make-comeback-despite-tough-economy⟩. Retrieved on August 31, 2011.

9. Dufey, A. (2006). *Biofuels production, trade and sustainable development: emerging issues* (No. 2). Iied.
10. Heaton, E. A., Schulte, L. A., Berti, M., Langeveld, H., Zegada-Lizarazu, W., Parrish, D., & Monti, A. (2013). Managing a second-generation crop portfolio through sustainable intensification: Examples from the USA and the EU. *Biofuels, Bioproducts and Biorefining, 7*(6), 702–714.
11. Diario oficial de la federación. http://www.dof.gob.mx/nota_detalle.php?codigo=5450011&fecha=29/08/2016
12. Renewable Fuels Association. 2021 Ethanol Industry Outlook. Available online: https://ethanolrfa.org/wp-content/uploads/2021/02/RFA_Outlook_2021_fin_low.pdf/. Accessed on September 22, 2021.
13. Renewable Fuels Association. 2020 Ethanol Industry Outlook. Available online: https://ethanolrfa.org/file/21/2020-OutlookFinal-for-Website.pdf/. Accessed on October 12, 2021
14. New Energy Blue to Construct Cellulosic Biorefinery in Iowa. Available online: http://biomassmagazine.com/articles/18180/new-energy-blue-to-construct-cellulosic-biorefinery-in-iowa/. Accessed on 12 October 2021.
15. Renewable Fuels Association Analysis of Public and Private Data. Available online: https://ethanolrfa.org/markets-andstatistics/annual-ethanol-production. Accessed on December 1, 2021.
16. USDA Foreign Agricultural Service. Vietnam Ethanol Background Report. 11 August 2020. Available online: https://www.fas.usda.gov/data/vietnam-vietnam-ethanol-background-report. Accessed on May 18, 2021.
17. India Biofuels Annual Report 2020, Global Agricultural Information Network; GAIN Report Number IN2021-0072; Foreign Agricultural Service, U.S. Department of Agriculture: Washington, DC, USA.
18. İçöz, E., Tuğrul, K. M., Saral, A., & İçöz, E. (2009). Research on ethanol production and use from sugar beet in Turkey. *Biomass and Bioenergy, 33*(1), 1–7.
19. Zabed, H., Faruq, G., Sahu, J. N., Azirun, M. S., Hashim, R., & Nasrulhaq Boyce, A. (2014). Bioethanol production from fermentable sugar juice. *The Scientific World Journal, 2014*.
20. Ethanol Fuel Basics, US DOE, Alternative Fuels Data Center [Internet]. Available from: https://afdc.energy.gov/fuels/ethanol_fuel_basics.Html
21. Balat, M., Balat, H., & Öz, C. (2008). Progress in bioethanol processing. *Progress in Energy and Combustion Science, 34*(5), 551–573.
22. Linoj, K. N. V., Prabha, D., Anandajit, G., & Sameer, M. (2006). Liquid biofuels in South Asia: Resources and technologies. *Asian Biotechnology and Development Review, 8*(2), 31–49.
23. Isikgor, F. H., & Becer, C. R. (2015). Lignocellulosic biomass: A sustainable platform for the production of bio-based chemicals and polymers. *Polymer Chemistry, 6*(25), 4497–4559.
24. Dashtban, M., Schraft, H., & Qin, W. (2009). Fungal bioconversion of lignocellulosic residues; opportunities & perspectives. *International journal of biological sciences, 5*(6), 578.
25. Haq, F., Ali, H., Shuaib, M., Badshah, M., Hassan, S. W., Munis, M. F. H., & Chaudhary, H. J. (2016). Recent progress in bioethanol production from lignocellulosic materials: A review. *International Journal of Green Energy, 13*(14), 1413–1441.
26. Peralta-Yahya, P. P., & Keasling, J. D. (2010). Advanced biofuel production in microbes. *Biotechnology Journal, 5*(2), 147–162.
27. Singh, R., Srivastava, M., & Shukla, A. (2016). Environmental sustainability of bioethanol production from rice straw in India: A review. *Renewable and Sustainable Energy Reviews, 54*, 202–216.
28. Saini, J. K., Saini, R., & Tewari, L. (2015). Lignocellulosic agriculture wastes as biomass feedstocks for second-generation bioethanol production: concepts and recent developments. *3 Biotech, 5*(4), 337–353.
29. Behera, S., Singh, R., Arora, R., Sharma, N. K., Shukla, M., & Kumar, S. (2015). Scope of algae as third generation biofuels. *Frontiers in bioengineering and biotechnology, 2*, 90.
30. Sanchez, O. J., & Cardona, C. A. (2008). Trends in biotechnological production of fuel ethanol from different feedstocks. *Bioresource Technology, 99*(13), 5270–5295.

31. Jambo, S. A., Abdulla, R., Azhar, S. H. M., Marbawi, H., Gansau, J. A., & Ravindra, P. (2016). A review on third generation bioethanol feedstock. *Renewable and sustainable energy reviews, 65*, 756–769.
32. Talebnia, F. (2015). Bioethanol from lignocellulosic wastes: current status and future prospects. In *Lignocellulose-based bioproducts* (pp. 175–206). Springer
33. Bušić, A., Marđetko, N., Kundas, S., Morzak, G., Belskaya, H., Ivančić Šantek, M., Komes, D., Novak, S., & Šantek, B. (2018). Bioethanol production from renewable raw materials and its separation and purification: A review. *Food Technology and Biotechnology, 56*(3), 289–311.
34. Özçimen, D., İnan, B., & Biernat, K. (2015). An overview of bioethanol production from algae. *Biofuels-Status and Perspective*, 141–162.
35. Metsoviti, M. N., Papapolymerou, G., Karapanagiotidis, I. T., & Katsoulas, N. (2019). Comparison of growth rate and nutrient content of five microalgae species cultivated in greenhouses. *Plants, 8*(8), 279.
36. Sankaran, R., Show, P. L., Nagarajan, D., & Chang, J. S. (2018). Exploitation and biorefinery of microalgae. In *Waste Biorefinery* (pp. 571–601). Elsevier.
37. Chen, C. Y., Zhao, X. Q., Yen, H. W., Ho, S. H., Cheng, C. L., Lee, D. J., Bai, F. W., & Chang, J. S. (2013). Microalgae-based carbohydrates for biofuel production. *Biochemical Engineering Journal, 78*, 1–10.
38. Sun, Y., & Cheng, J. (2002). Hydrolysis of lignocellulosic materials for ethanol production: A review. *Bioresource Technology, 83*(1), 1–11.
39. Yusuf, A. A., & Inambao, F. L. (2021). Effect of low bioethanol fraction on emissions, performance, and combustion behavior in a modernized electronic fuel injection engine. *Biomass Conversion and Biorefinery, 11*(3), 885–893.
40. https://www.globenewswire.com/news-release/2020/03/26/2007160/0/en/Global-Hand-Sanitizer-Market-Is-Expected-to-Reach-USD-2-14-Billion-by-2027-Fior-Markets.html
41. http://www.prnewswire.com/newsreleases/red-river-biorefinery-nowproducingpharmaceutical-grade-usp-ethanol-for-hand-sanitizers-anddisinfectants-301146352.html
42. Bajpai, P. (2007). Bioethanol. PIRA Technology Report, Smithers PIRA UK.
43. EPA (Environmental Protection Agency). (1997). Fact sheet: EPA's revised ozone standard, United States. www.epa.gov/ttn/oarpg/naaqsfin/o3fact.html. Accessed March 2001.
44. Hu, Z., Pu, G., Fang, F., & Wang, C. (2004). Economics, environment, and energy life cycle assessment of automobiles fueled by bio-ethanol blends in China. *Renewable Energy, 29*(14), 2183–2192.
45. Department of Energy. (2006). Bioethanol feedstocks. https://www1.eere.energy.gov/biomass/abcs_biofuels.html
46. Department of Energy. (2006). *Guidebook for handling, storing, & dispensing fuel ethanol prepared for the U.S. department of energy by the center for transportation research energy systems division.* Argonne National Laboratory.
47. Lapuerta, M., Armas, O., & Garcia-Contreras, R. (2007). Stability of diesel–bioethanol blends for use in diesel engines. *Fuel, 86*(10–11), 1351–1357.
48. Jia, L. W., Shen, M. Q., Wang, J., & Lin, M. Q. (2005). Influence of ethanol–gasoline blended fuel on emission characteristics from a four-stroke motorcycle engine. *Journal of Hazardous Materials, 123*(1–3), 29–34.
49. Concepts, E.B. (2017). *European technology and innovation platform bioenergy.* Gülzow
50. Flores-Gómez, C. A., Escamilla Silva, E. M., Zhong, C., Dale, B. E., da Costa Sousa, L., & Balan, V. (2018). Conversion of lignocellulosic agave residues into liquid biofuels using an AFEX™-based biorefinery. *Biotechnology for Biofuels, 11*(1), 1–18.
51. Kazi, F. K., Fortman, J. A., Anex, R. P., Hsu, D. D., Aden, A., Dutta, A., & Kothandaraman, G. (2010). Techno-economic comparison of process technologies for biochemical ethanol production from corn stover. *Fuel, 89*, S20–S28.
52. Fedorova, E., Caló, A., & Pongrácz, E. (2019). Balancing socio-efficiency and resilience of energy provisioning on a regional level, case Oulun Energia in Finland. *Clean Technologies, 1*(1), 273–293.

53. Rass-Hansen, J., Falsig, H., Jørgensen, B., & Christensen, C. H. (2007). Bioethanol: Fuel or feedstock? *Journal of Chemical Technology & Biotechnology: International Research in Process, Environmental & Clean Technology, 82*(4), 329–333.
54. Schnell, P. G., & Akin, C. (1979). Functional properties of yeast grown on ethyl alcohol. *Journal of the American Oil Chemists' Society, 56*(1), A82–A85.
55. Oki, T., Sayama, Y., Nishimura, Y., & Ozaki, A. (1968). L-Glutamic acid formation by microorganisms from ethanol. *Agricultural and Biological Chemistry, 32*(1), 119–120.
56. Nakayama, K., & Araki, K. (1973). *U.S. Patent No. 3,708,395.* Washington, DC: U.S. Patent and Trademark Office.
57. Hideaki, Y., Kazuoki, O., Terukazu, N. and Yoshihiro, T., 1980. Production of L-amino acids. *US Patent* 4276380.
58. Giudici, P., Gullo, M., & Solieri, L. (2009). Traditional balsamic vinegar. In *Vinegars of the world* (pp. 157–177). Springer, Milano.
59. Tiwari, J. N., Tiwari, R. N., & Kim, K. S. (2012). Zero-dimensional, one-dimensional, two-dimensional and three-dimensional nanostructured materials for advanced electrochemical energy devices. *Progress in Materials Science, 57*(4), 724–803.
60. Kodama, R. H. (1999). Magnetic nanoparticles. *Journal of Magnetism and Magnetic Materials, 200*(1–3), 359–372.
61. Laurent, S., Forge, D., Port, M., Roch, A., Robic, C., Vander Elst, L., & Muller, R. N. (2008). Magnetic iron oxide nanoparticles: Synthesis, stabilization, vectorization, physicochemical characterizations, and biological applications. *Chemical Reviews, 108*(6), 2064–2110.
62. Alftrén, J., & Hobley, T. J. (2013). Covalent immobilization of β-glucosidase on magnetic particles for lignocellulose hydrolysis. *Applied Biochemistry and Biotechnology, 169*(7), 2076–2087.
63. Kim, Y. K., & Lee, H. (2016). Use of magnetic nanoparticles to enhance bioethanol production in syngas fermentation. *Bioresource Technology, 204*, 139–144.
64. Han, H., Cui, M., Wei, L., Yang, H., & Shen, J. (2011). Enhancement effect of hematite nanoparticles on fermentative hydrogen production. *Bioresource Technology, 102*(17), 7903–7909.
65. El-Kemary, M., Nagy, N., & El-Mehasseb, I. (2013). Nickel oxide nanoparticles: Synthesis and spectral studies of interactions with glucose. *Materials Science in Semiconductor Processing, 16*(6), 1747–1752.
66. Sanusi, I. A., Faloye, F. D., & Gueguim Kana, E. B. (2019). Impact of various metallic oxide nanoparticles on ethanol production by Saccharomyces cerevisiae BY4743: Screening, kinetic study and validation on potato waste. *Catalysis Letters, 149*(7), 2015–2031.
67. Kuhad, R. C., Gupta, R., Khasa, Y. P., & Singh, A. (2010). Bioethanol production from Lantana camara (red sage): Pretreatment, saccharification and fermentation. *Bioresource Technology, 101*(21), 8348–8354.
68. Satyanagalakshmi, K., Sindhu, R., Binod, P., Janu, K. U., Sukumaran, R. K., & Pandey, A. (2011). Bioethanol production from acid pretreated water hyacinth by separate hydrolysis and fermentation.
69. Prasertwasu, S., Khumsupan, D., Komolwanich, T., Chaisuwan, T., Luengnaruemitchai, A., & Wongkasemjit, S. (2014). Efficient process for ethanol production from Thai Mission grass (*Pennisetum polystachion*). *Bioresource Technology, 163*, 152–159.
70. Mishima, D., Kuniki, M., Sei, K., Soda, S., Ike, M., & Fujita, M. (2008). Ethanol production from candidate energy crops: water hyacinth (*Eichhornia crassipes*) and water lettuce (*Pistia stratiotes* L.). *Bioresource Technology, 99*(7), 2495–2500.
71. Meng, F., Dornau, A., Mason, S. J. M., Thomas, G. H., Conradie, A., & McKechnie, J. (2021). Bioethanol from autoclaved municipal solid waste: Assessment of environmental and financial viability under policy contexts. *Applied Energy, 298*, 117118.
72. Vaez, S., Karimi, K., Mirmohamadsadeghi, S., & Jeihanipour, A. (2021). An optimal biorefinery development for pectin and biofuels production from orange wastes without enzyme consumption. *Process Safety and Environmental Protection, 152*, 513–526.

73. Wang, Z., Lv, J., Gu, F., Yang, J., & Guo, J. (2020). Environmental and economic performance of an integrated municipal solid waste treatment: A Chinese case study. *Science of the Total Environment, 709*, 136096.
74. Sivamani, S., Chandrasekaran, A. P., Balajii, M., Shanmugaprakash, M., Hosseini-Bandegharaei, A., & Baskar, R. (2018). Evaluation of the potential of cassava-based residues for biofuels production. *Reviews in Environmental Science and Bio/Technology, 17*(3), 553–570.

Chapter 2
Understanding of Different Processing Technologies for Bioethanol Production

Abstract Bioethanol has the potential to address the energy crisis by replacing fossil fuels. The need for energy is increasing day by day as a result of the rising population and industrialization. Simultaneously, global bioethanol production is steadily increasing. Bioethanol production's environmental consequences are entirely reliant on substrate accessibility and conversion technology. Pretreatment, enzymatic hydrolysis, sugar fermentation, and distillation are four steps that go into making bioethanol from agro-residues. Several hurdles and limits face these operations, including biomass transport and handling, as well as an effective pretreatment technique for separating lignin from lignocellulosic biomass residues. This chapter examines the present status of different bioethanol production technologies such as Separate Hydrolysis and Fermentation, Simultaneous Saccharification and Fermentation, Simultaneous Saccharification and Co-fermentation, and Consolidated Bioprocessing.

Keywords Bioethanol · Lignocellulose biomass · Processing technologies

2.1 Introduction

The rising standards of life, economic and industrial expansion, and population increase have all posed severe difficulties for India's energy sector. Even though the country has one of the world's fastest-expanding economies, the fundamental energy demands of tens of millions of its residents have yet to be met. To meet the basic energy needs of its inhabitants, power generation capacity must expand to about 800 GW by 2031–2032, up from the existing capacity of around 183 GW, including all captive facilities [1]. An energy shortage has become a worldwide problem due to rising global energy demand, which is estimated to increase by 50% by 2025. Concerns about energy security, the depletion of fossil resources, and global warming have prompted scientists to search for alternative, sustainable, and renewable energy systems. Lignocellulosic biomass has been discovered to be abundant, renewable, and cost-effective, and it might be used to make biofuels and other high-value compounds [2]. Depending on the type of raw material, bioethanol is usually divided into three categories. The 1st is bioethanol, which is made from sugar-based

N. K. Aggarwal et al., *Bioethanol Production*, Green Chemistry and Sustainable Technology, https://doi.org/10.1007/978-3-031-05091-6_2

plants like sugarcane and sugar beet. The 2nd is made from starch-based materials, including maize and wheat grains, as well as root and tuber crops. Cellulosic bioethanol, also known as 2G bioethanol, is generated from cellulosic materials such as crop residue (for example, rice straw and maize stover) and woody materials. The US and Brazil are the world's top producers of bioethanol, making 84% of the world's total. Europe placed in third place with 5443 million liters of ethanol in 2019 (5% of global output). Fuel ethanol is currently made from food crops such as sugar cane, corn, wheat, and barley. This process is known as 1G bioethanol production, and the technology has reached industrial maturity. In 2019, 94% of the ethanol produced in the United States was derived from corn starch. Brazil, on the other hand, gets almost all of its bioethanol from sugarcane (99% of total production). Bioethanol is generated in the European Union from a variety of feedstocks, including sugar beet (57.9%), maize (19.6%), wheat (18.7%), barley (1.9%), and rye (1.9%) [3]. An acid and/or enzyme-based fermentative process can be used to convert biomass to products such as fuel bioethanol. This is one of the viable future technologies for utilizing this abundant lignocellulose [4]. The production pathways have a significant impact on the quality of bioethanol generated. Because bioethanol production, in general, comprises many sequential techniques, such as pretreatment, hydrolysis, fermentation, and distillation, each stage is branched, and each branch will produce ethanol of varying quality and cost. Furthermore, current accessible technologies expand the range of bioethanol production pathways. Acid hydrolysis by H_2SO_4, for example, yields a sugar conversion efficiency of 76% for cellulose to C_6 and 90% for hemicellulose to C_5 [5]. The use of lignocellulosic material for ethanol production requires four key processes:

1. Pretreatment of lignocellulosic biomass

Method of pretreatment

- Microwave, ultrasonication, and extrusion are examples of physical processes.
- Alkaline and acid hydrolysis is two types of chemical hydrolysis.
- Physicochemical: Steam explosions, AFEX.
- Biological: Enzymatic pretreatment, whole-cell pretreatment.

2. Enzymatic saccharification of pretreated dry biomass to liberate fermentable sugars.
3. Fermentation: Hydrolyzed sugars are fermented with the help of fermenting organisms into bioethanol.
4. Distillation.

2.2 Feedstocks for the Production of Bioethanol

The three polymers, cellulose, hemicellulose, and lignin, as well as proteins and ash, make up the majority of lignocellulosic biomass. Lignocellulosic biomass major

Table 2.1 Chemical composition of various feedstocks

Biomass	Cellulose	Hemicellulose	Lignin	References
Eucalyptus	54	18	21	[6]
Miscanthus	41–53	20–25.3	20–23.8	[7]
Napier grass	47	31	22	[8]
Poplar	42–49	16–23	21–29	[9]
Bamboo	45	24	20	[10]
Corn cob	43.7	23.7	12.5	[11]
Corn stover	38	23	20	[12]
Rice straw	38	32	12	[13]
Barley straw	33	26	19	[14]

raw material uses for bioethanol production. The chemical composition of lignocellulosic materials varies not only in terms of the amount of non-cellulosic carbohydrates present but also in terms of their kind and the presence of proteins or phenolic chemicals. Lignin is tightly bound to cellulose or hemicelluloses, leaving the polysaccharides inaccessible to cellulases or hemicellulases for digestion. The world's largest source of natural carbohydrates is lignocellulosic biomass. Because they are composed of cellulose fibers surrounded by an amorphous matrix of hemicelluloses and lignin, the complexity of converting LCB into liquid fuel is linked to its recalcitrance. Cellulose is a structural part of lignocellulosic biomass or the glucose polymer, which is linked to reduced or non-reduced ends via a β-(1–4) glycosidic interaction. Hemicellulose is an amorphous, branched polymer that has at least 500–3000 sugar monomers in it. It is found in the primary cell wall. Lignin is a heteropolymer that gives crops their rigidity and forms a complex network within cellulose microfibrils. Holocellulases can break down cellulose or hemicellulose into C_5 and C_6, which can then be fermented into ethanol. Lignin, on the other hand, is a polyphenolic mixture that cannot be fermented. Table 2.1 shows the chemical composition of various lignocellulosic biomass utilized for ethanol production.

2.3 Different Processing Technologies for Bioethanol Production

Various technologies for producing bioethanol from LCB have been developed by scientists all over the world. These processes include the following:

- SHF: Separate Hydrolysis and Fermentation
- SSF: Simultaneous Saccharification and Fermentation
- SSCF: Simultaneous Saccharification and Co-fermentation
- CBP: Consolidated Bioprocessing.

2.3.1 Concept of Separate Hydrolysis and Fermentation (SHF) Process

This technique is the most widely utilized because of the versatility it provides in respect of hydrolysis process selection. SHF is a conventional design in which the cellulose is hydrolyzed after pretreatment of the biomass and then fermented. After that, the hydrolysate is neutralized and the insoluble and solid fractions are separated. It is then fermented to produce alcohol. The insoluble fraction is kept and processed with glycosidase and cellulase to liberate hexose sugar, which is subsequently fermented to make bioethanol. Lignin is burned as a residue insoluble material to generate energy for the overall process [15, 16]. In this process, an externally generated enzyme is employed to hydrolyze pretreated LCB to yield sugar monomers. Fermenting bacteria employ the resultant enzymatic hydrolysate to produce biofuels. Because the temperature optima of hydrolytic enzymes (about 50 °C) and fermentation (30–35 °C) are different, all processes are carried out separately. Even though both processes used maize cob biomass and the same amount of *Trichoderma reesei* enzymes, the SHF process made a little more ethanol than the SSF process did (76.47%), but both processes made a lot of bioethanol (Fig. 2.1).

In a study on *G. tenuistipitata*, acid hydrolysis was used to produce the reducing sugars, which were then fermented with *S. cerevisiae* to yield a bioethanol yield of 0.042 g/g reducing sugars [18].

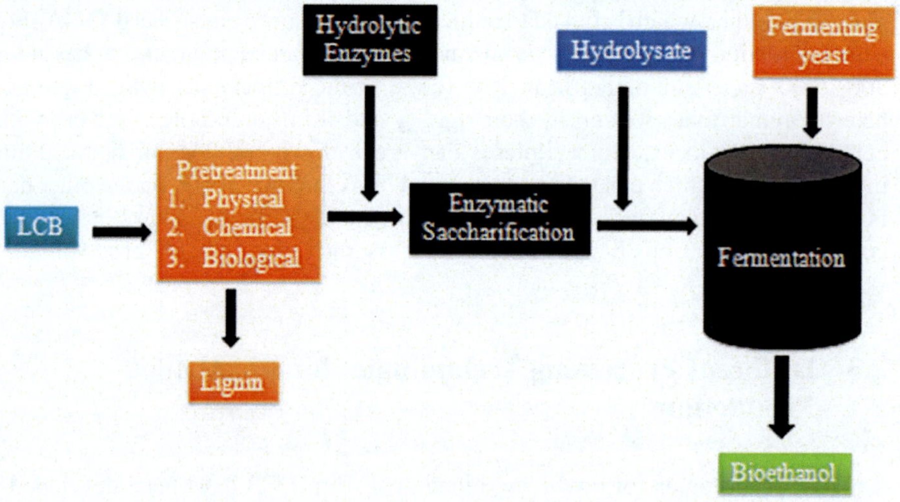

Fig. 2.1 Flow of separate hydrolysis and fermentation processes

2.3.2 *Concept of Simultaneous Saccharification and Fermentation (SSF) Process*

Gauss et al. proposed the concept of doing enzymatic hydrolysis and fermentation at the same time in a patent from 1976 [19]. The pretreated biomass is enzymatically saccharified and the enzymatic hydrolysate is fermented in the same vessel using this process. The process parameters for optimizing both the enzymes used for saccharification and the microbes used for fermentation at the same time are the most important difficulties with this technique. The glucose generated during hydrolysis is utilized by the yeast simultaneously, eliminating its accumulation in the reaction media and, as a result, reducing cellulase inhibition by hydrolysis products, cellobiose, and glucose [20]. According to Ohgren et al. [21], the SSF procedure yielded 13% more LC ethanol than the SHF technique. Cold-adaptive hydrolytic enzymes and thermophilic yeast are mostly used in this technique. It is carried out in a single vessel at a temperature of around 40 °C.

Kluyveromyces marxianus is the most promising of them, as different strains of this yeast can thrive at high temperatures (45–52 °C) and produce ethanol at temperatures ranging from 38 to 45 °C. *K. marxianus* also has the benefit of a rapid growth rate and the capacity to use a variety of sugar sources at high temperatures, including galactose, xylose arabinose, or mannose [16, 22] (Fig. 2.2).

SSF has many benefits:

- It increases the hydrolysis rate by converting sugars that inhibit cellulase activity at the same time.
- It reduces the need for enzymes.

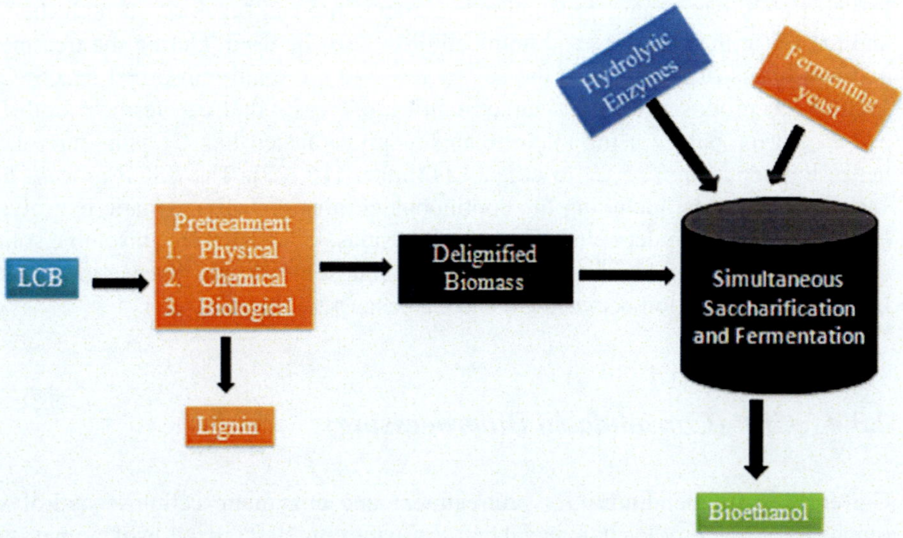

Fig. 2.2 Flow of Simultaneous saccharification and fermentation processes

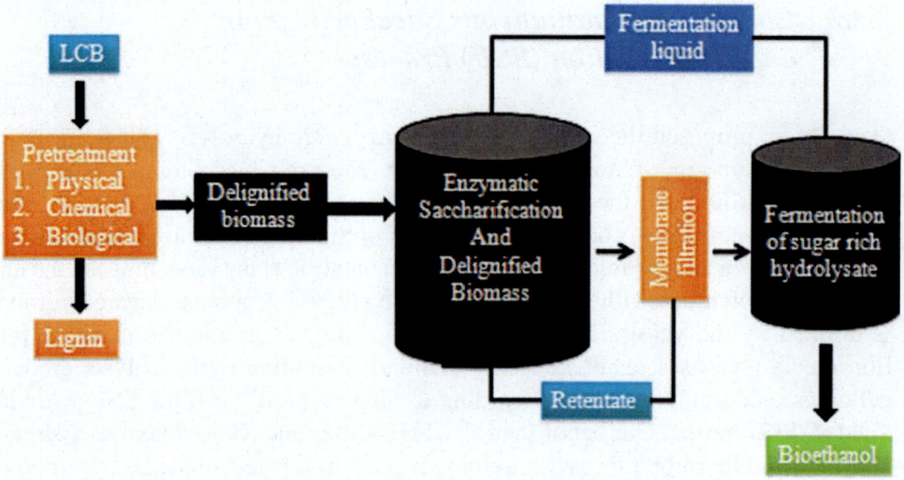

Fig. 2.3 Flow of the Simultaneous saccharification and co-fermentation process

- It increases yield.
- It reduces total process time.
- Contamination risk is decreased.

2.3.3 Concept of Simultaneous Saccharification and Co-fermentation (SSCF) Process

For cost-effective ethanol production, microbes that can break down both the C6 and the C5 in the cellulose and hemicellulose must be used. During the treatment and hydrolysis of LCB, all of the sugars released are totally absorbed into SSCF. It has been proposed to use a mixed culture of yeasts that can ferment both C_5 and C_6 sugars, but C_6 using microorganisms grows faster than C_5 using microbes, hence the conversion of C_6 to ethanol is higher [16, 23]. The SSCF process has various advantages, including the continuous elimination of enzymatic hydrolysis end-products that hinder cellulases or β-glucosidases. A high water-insoluble solids content helps mix and make more bioethanol, but it also keeps glucose low, which makes co-fermentation between C6 and C5 easier [24, 25] (Fig. 2.3).

2.3.4 CBP (Consolidated Bioprocessing)

Currently, the expenditures for pretreatment and enzymatic cellulose hydrolysis remain the most significant financial barriers to the commercialization of biomass-to-ethanol conversion technology. Consolidated bioprocess is a completely integrated process that includes enzyme synthesis, biomass digestion, and glucose fermentation all in one step or a single reactor. The consolidated bioprocessing is a process-process fusion for feedstock conversion into ethanol.

Two alternative approaches have been used to equip the microorganism with the qualities required, including a high hydrolysis rate, resistance to a chemical formed from pretreatment, and correct fermentation, in the search for a CBP functioning organism:

Native technique: The native method of increasing ethanol productivity looks at organisms that have the natural ability to synthesize multiple enzymes and employ a variety of substrates. The native strains recommended for use in consolidated bioprocess are mostly wild-type strains with little data about them. Only a handful of them have genetic manipulation tools, and only a few of their metabolisms have been thoroughly examined. The majority of genetic engineering efforts in cellulolytic fungus are aimed at enhancing cellulase production. However, there is an increasing interest in using these organisms to generate biofuels. Fungi, bacteria that produce enzymes, and cellulosome-forming bacteria are the three types of options for the native strategy [26, 27].

Recombinant technique: The recombinant technique aims to enhance the hydrolytic capacity of fermenting microbes. As a result, it is important to choose host species with the appropriate properties before engineering microorganisms for biofuels production, with a focus on strains that can use low-cost substrates, are resistant to environmental stress, and produce a high yield of the intended product [28, 29] (Fig. 2.4).

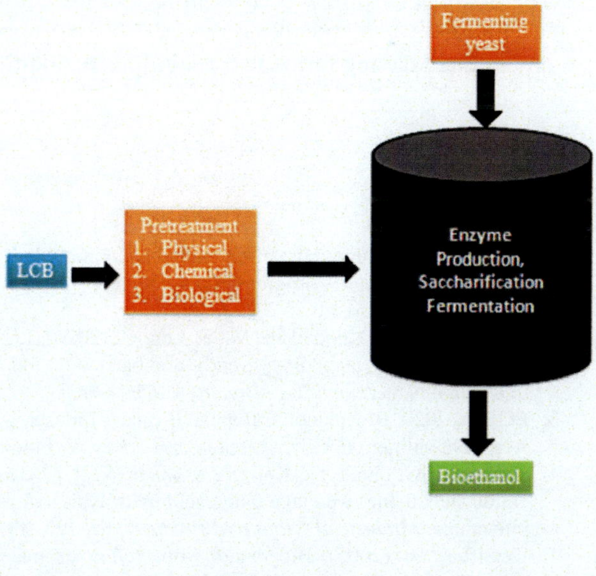

Fig. 2.4 Flow of Consolidated Bioprocess

2.4 Summary

Bioethanol production from LCB has a huge potential. Due to the abundance of lignocellulosic biomass resources and its potential as a cleaner and ecologically friendlier biofuels, bioethanol synthesis from 2G feedstocks has gotten a lot of attention. Like any other process, SSF faces the challenge of achieving the highest degree of hydrolysis and bioethanol output feasible. More efficient C_5 fermenting yeasts with improved hydrolysate robustness, as well as more efficient enzymes and enzyme combinations, will surely continue to be developed. The discovery of fermenting microorganisms capable of converting $C_{6,}$ C_5 and sugar alcohols from biomass to bioethanol, rather than the choice between SHF and SSF methods, limited the effectiveness of the fermentation stage. There are now only minor variations in ethanol yields between SHF and SSF. SSCF has been identified as a viable method for producing ethanol from lignocellulosic materials high in C_5. When researchers first started working on valuing LCB, several bioprocessing techniques, such as SSF, SHF, SSCF as well as CBP, emerged as essential to the environment and the economy. Microbes are being genetically modified to perform all tasks, with or without pretreatment, by researchers. Because CBP is a one-process technology using less equipment or a single reactor, it saves time and money in biorefineries. Because some microbes can convert LCB to bioethanol and other useful materials in a single cycle or reactor with or without pretreatment, in the future, CBP will play a key role in converting LCB to bioethanol and other essential products in a single process or reactor with or without pretreatment, making it more cost-effective and time-saving. Emerging technologies, such as CBP, are promising because they decrease operation stages and chemical inhibitors, and they can be improved by lignin, which is an energy-self-sustaining co-product. These processes are commonly linked with thermophilic and cellulolytic bacteria, like *T. reesei, P. chrysosporium*, and *C. cellulolyticum*, with some of them having fermentative abilities in addition to hydrolytic abilities.

References

1. India. (2011). Biomass for sustainable development-lessons for decentralized energy delivery village energy security programme. *Document of the World Bank*. http://www.mnre.gov.in/pdf/VESP-Final-Report-July%202011.pdf
2. Su, T., Zhao, D., Khodadadi, M., & Len, C. (2020). Lignocellulosic biomass for bioethanol: Recent advances, technology trends, and barriers to industrial development. *Current Opinion in Green and Sustainable Chemistry, 24*, 56–60.
3. EUBIA. 2020. Bioethanol; European Biomass Industry Association: *Brussels*, Belgium, 2020; Available online: https://www.eubia.org/ cms/wiki-biomass/biofuels/bioethanol
4. Keshav, P. K., Shaik, N., Koti, S., & Linga, V. R. (2016). Bioconversion of alkali delignified cotton stalk using two-stage dilute acid hydrolysis and fermentation of detoxified hydrolysate into ethanol. *Industrial Crops and Products, 91*, 323–331.
5. Demirbaş, A. (2005). Bioethanol from cellulosic materials: a renewable motor fuel from biomass. *Energy Sources, 27*(4), 327–337.

6. Isikgor, F. H., & Becer, C. R. (2015). Lignocellulosic biomass: a sustainable platform for the production of bio-based chemicals and polymers. *Polymer Chemistry, 6*(25), 4497–4559.
7. Gismatulina, Y. A., Budaeva, V. V., & Sakovich, G. V. (2018). Nitrocellulose synthesis from miscanthus cellulose. *Propellants, Explosives, Pyrotechnics, 43*(1), 96–100.
8. Reddy, K. O., Maheswari, C. U., Dhlamini, M. S., Mothudi, B. M., Kommula, V. P., Zhang, J., Zhang, J., & Rajulu, A. V. (2018). Extraction and characterization of cellulose single fibers from native african napier grass. *Carbohydrate Polymers, 188*, 85–91.
9. Sannigrahi, P., Ragauskas, A. J., & Tuskan, G. A. (2010). Poplar as a feedstock for biofuels: A review of compositional characteristics. *Biofuels, Bioproducts and Biorefining, 4*(2), 209–226.
10. Li, X., Sun, C., Zhou, B., & He, Y. (2015). Determination of hemicellulose, cellulose and lignin in moso bamboo by near infrared spectroscopy. *Scientific Reports, 5*(1), 1–11.
11. Saliu, B. K., & Sani, A. (2012). Bioethanol potentials of corn cob hydrolysed using cellulases of *Aspergillus niger* and *Penicillium decumbens. Excli Journal, 11*, 468.
12. Wan, C., & Li, Y. (2010). Microbial pretreatment of corn stover with *Ceriporiopsis subvermispora* for enzymatic hydrolysis and ethanol production. *Bioresource Technology, 101*(16), 6398–6403.
13. Lu, P., & Hsieh, Y. L. (2012). Preparation and characterization of cellulose nanocrystals from rice straw. *Carbohydrate Polymers, 87*(1), 564–573.
14. Duque, A., Doménech, P., Álvarez, C., Ballesteros, M., & Manzanares, P. (2020). Study of the bioprocess conditions to produce bioethanol from barley straw pretreated by combined soda and enzyme-catalyzed extrusion. *Renewable Energy, 158*, 263–270.
15. Huber, G. W., Iborra, S., & Corma, A. (2006). Synthesis of transportation fuels from biomass: Chemistry, catalysts, and engineering. *Chemical Reviews, 106*(9), 4044–4098.
16. Vohra, M., Manwar, J., Manmode, R., Padgilwar, S., & Patil, S. (2014). Bioethanol production: Feedstock and current technologies. *Journal of Environmental Chemical Engineering, 2*(1), 573–584.
17. Li, Y. H., Zhang, X. Y., Zhang, F., Peng, L. C., Zhang, D. B., Kondo, A., Bai, F. W., & Zhao, X. Q. (2018). Optimization of cellulolytic enzyme components through engineering *Trichoderma reesei* and on-site fermentation using the soluble inducer for cellulosic ethanol production from corn stover. *Biotechnology for Biofuels, 11*(1), 1–14.
18. Chirapart, A., Praiboon, J., Puangsombat, P., Pattanapon, C., & Nunraksa, N. (2014). Chemical composition and ethanol production potential of Thai seaweed species. *Journal of Applied Phycology, 26*(2), 979–986.
19. Gauss, W. F., Suzuki, S., Takagi, M., & Bio Research Center Co Ltd. (1976). *Manufacture of alcohol from cellulosic materials using plural ferments.* U.S. Patent 3,990,944.
20. Olofsson, K., Bertilsson, M., & Lidén, G. (2008). A short review on SSF—An interesting process option for ethanol production from lignocellulosic feedstocks. *Biotechnology for Biofuels, 1*(1), 1–14.
21. Öhgren, K., Bura, R., Lesnicki, G., Saddler, J., & Zacchi, G. (2007). A comparison between simultaneous saccharification and fermentation and separate hydrolysis and fermentation using steam-pretreated corn stover. *Process Biochemistry, 42*(5), 834–839.
22. Cardona, C. A., & Sánchez, Ó. J. (2007). Fuel ethanol production: Process design trends and integration opportunities. *Bioresource Technology, 98*(12), 2415–2457.
23. Sanchez, O. J., & Cardona, C. A. (2008). Trends in biotechnological production of fuel ethanol from different feedstocks. *Bioresource Technology, 99*(13), 5270–5295.
24. Zhang, M., Wang, F., Su, R., Qi, W., & He, Z. (2010). Ethanol production from high dry matter corncob using fed-batch simultaneous saccharification and fermentation after combined pretreatment. *Bioresource Technology, 101*(13), 4959–4964.
25. Öhgren, K., Bengtsson, O., Gorwa-Grauslund, M. F., Galbe, M., Hahn-Hägerdal, B., & Zacchi, G. (2006). Simultaneous saccharification and co-fermentation of glucose and xylose in steam-pretreated corn stover at high fiber content with *Saccharomyces cerevisiae* TMB3400. *Journal of biotechnology, 126*(4), 488–498.
26. Zuroff, T. R., Xiques, S. B., & Curtis, W. R. (2013). Consortia-mediated bioprocessing of cellulose to ethanol with a symbiotic *Clostridium phytofermentans*/yeast co-culture. *Biotechnology for Biofuels, 6*(1), 1–12.

27. Nagarajan, D., Lee, D. J., & Chang, J. S. (2019). Recent insights into consolidated bioprocessing for lignocellulosic biohydrogen production. *International Journal of Hydrogen Energy, 44*(28), 14362–14379.
28. Fan, Z. (2014). Consolidated bioprocessing for ethanol production. In *Biorefineries* (pp. 141–160). Elsevier.
29. Ábrego, U., Chen, Z., & Wan, C. (2017). Consolidated bioprocessing systems for cellulosic biofuel production. *Advances in Bioenergy, 2*, 143–182.

Chapter 3
Current Trends in Pretreatment Technologies for Bioethanol Production: Biorefinery Concept

Abstract The call for an alternative and clean energy has grown as fossil sources have been depleted and energy needs have increased. The use of readily available lignocellulosic biomass for creating cost-effective and environmentally beneficial large-scale biorefinery applications has given this field a much-needed boost. Pretreatment is an important step in converting biomass into high-value products like sugars and biofuels. To overcome the recalcitrance of lignocellulosic biomass and accelerate its decomposition into individual units—like, cellulose, hemicellulose, and lignin-different pretreatment techniques are used. Traditional pretreatment procedures are unsuitable for industrial scale-up due to their lack of sustainability and practicability. Milling, microwave, extrusion, ammonia fiber explosion, solvents, and other selected physical and chemical pretreatment techniques are included in this paper.

Keywords Pretreatment technologies · Biorefinery · Lignocellulose · Energy · Economy

3.1 Introduction

Energy is a vital aspect that influences a country's socioeconomic progress. According to the Global Status Report (GSR) on energy, non-renewable fossil fuels like coal, petroleum, and natural gases account for 78.4% of total energy consumption, while renewable alternatives, for example, hydropower, solar, wind, and biomass account for only 19% [1]. Fossil fuels are limited in supply and pollute the environment severely. Several attempts have been made to more effectively utilize fossil fuels in order to enhance their specific heat while emitting less pollutants into the environment. Microwave processing of coal enhanced the digestibility of high-ash coal and improve efficiency of combustion [2]. When a wide variety of brown coal to anthracite slurries was treated under method of electromagnetic pretreatment, combustion properties like time required for ignition, temperature for ignition, maximum flame temperature, and time–temperature correlations improved [3]. Biomass coal co-firing has also been implemented to reduce net carbon dioxide emissions [4]. In situ pretreatment of alumina aids in cobalt catalysts employed in the

N. K. Aggarwal et al., *Bioethanol Production*, Green Chemistry and Sustainable Technology, https://doi.org/10.1007/978-3-031-05091-6_3

Fischer–Tropsch synthesis, i.e., natural gas to liquid fuel that result in a reduction in the amount of long-chain hydrocarbons (C25+) such as waxes and an increase in the synthesis of liquid transport fuels like diesel and gasoline [5]. Though, there are also other restrictions connected with fossil fuels, such as limited reserves in politically unbalanced parts of the world, uncertain price swings, non-renewability, emission of greenhouse gases, and so on [6], which limit its usage for future energy and chemical demands. Different chemicals are made from fossil fuels, and the depletion of fossil fuel reserves is a major worry for several chemical businesses. To attain "healthy environment-healthy human being" and "socioeconomic prosperity," civilization is currently transitioning from the primordial paradigm of "take, make, and dispose of" to "reuse and recovery" of resources [7]. Plants, animals, and microorganisms provide the most sensible carbon-based feedstock: biomass. It has developed as a viable renewable bioresource that can be used in place of non-renewable sources. The predicted return is so large that economists, researchers, and state policymakers have invented the term "bio-based economy," which refers to a parallel economy derived from bio-based products and is also known as "circular bioeconomy" because of its renewable nature [8]. Lignocellulosic feedstocks (LCF) are plentiful among diverse feedstocks, with a yearly production of 200 109 tons per year. Many countries are becoming more interested in biorefining of LCB to synthesize biofuels and other useful chemicals to supplement fossil fuels [9–11]. LCB is made up of a variety of polymers, including polysaccharides like cellulose and hemicellulose, phenolic compound- lignin, as well as polar and non-polar components [9, 12, 13]. The complicated structure of LCF makes it resistant to enzymatic hydrolysis, making it difficult to convert it to biofuels. By disorganizing lignin and cleaving lignocellulosic biomass into its constituents, pretreatment aids in overcoming LCB's innate recalcitrance. Pretreatment improves the accessibility of hydrolyzing enzymes for cellulose and hemicellulose present in biomass, resulting in the production of hexose and pentose sugars, which are then further fermented to produce biofuels [14]. The hydrolysate of aromatic lignin and cellulosic hydrolysates is transformed into value-added compounds and forms the building units for economically useful chemicals [15]. Various pretreatment strategies have been developed with this in mind, and they can be classified as physical, chemical, physicochemical, or biological. The current research provides an overview of various biomass pretreatment procedures that are employed in laboratory, pilot, and large-scale biofuel production.

3.2 Composition of Lignocellulosic Biomass

To comprehend the structural complexity and resistant character of a typical LCB, a basic understanding of the constituent units present in biomass is required. The three principal components of LCB are cellulose, i.e., 35–50%, hemicellulose, 20–35%, and lignin 5–30%[16]. Depending on the source of the biomass, the percentage mix of the three separate components may vary substantially (Table 3.1). Table 3.1 shows that the cellulosic component is found in the highest percentage in most LCBs.

Table 3.1 Compositional analysis of different lignocellulosic feedstocks

Type of LCB	Cellulose (%)	Hemicellulose (%)	Lignin (%)	References
Sugarcane bagasse	35	35.8	16.1	[20]
Water hyacinth	24.5	34.1	8.6	[21]
Rice straw	35.8	21.5	24.4	[22]
Corn stalks	50	20	30	[23]
Beech wood	44.2	33.5	21.8	[24]
bamboo	41.8	18	29.3	[25]
Sugarcane tops	43	27	17	[26]
Hazelnut shell	25.2	28.2	42.1	[24]

Various glucose monomer units are linked linearly by −1,4 glycosidic bonds to form the polysaccharide cellulose [17]. The difficult and refractory nature of cellulose is due to its high degree of crystallinity in structure, higher polymerization approximately up to 10,000 units, and presence of complex network of inter- and intramolecularly hydrogen linked hydroxyl groups [16]. In comparison to the complex cellulosic fractions, hydrophilic and amorphous hemicellulose moieties in structure exhibit comparatively lower grades of polymerization. As a result, the hemicellulose fraction has little impact on the refractory nature of LCB. Hemicellulose is a heteropolymer with a branching structure. It is made up of sugars with five and six carbons that are connected together by −1,4 glycosidic bonds [18]. The hemicellulose component imparts stiffness to the overall biomass structure by acting as a binder between the cellulosic and lignin sections. Because both cellulose and hemicellulose contain reducing sugars, they are more important as a source of numerous economically viable compounds [19]. The lignin fraction, on the other hand, is composed of very intricate methoxylated phenylpropanoid units. It is mostly responsible for LCB's recalcitrance. Sinaphyl alcohol, p-coumaryl alcohol, coniferyl alcohol make up lignin, a three-dimensional macromolecule. These monolignols, also known as the S, G, and H units, are linked together by C–O–C and C–C links. The lignin portion can be catalyzed to produce various aromatics with additional value.

3.3 Pretreatment: A Necessity

Fractionation is the only option to make effective and comprehensive use of lignocellulosic feedstock due to its diverse composition. The complicated and highly packed structure of lignocellulose, on the other hand, makes it difficult to access and separate the various components so that they can be adequately valorized. Pretreatment, in this context, refers to a collection of actions aimed at breaking the carbohydrate–lignin matrix, allowing for easier access to and recovery of important chemicals.

3.4 Types of Pretreatment

Physical, chemical, physicochemical, and biological pretreatment methods have generally been categorized based on their driving force [27–29]. Because it was designed to be applied to unit processes, this categorization has become outdated [30]. The scope and aim of pretreatment have also altered. Following this trend, the current study begins a discussion of different kinds of pretreatment from various perspectives, utilizing nonexclusive categories.

3.4.1 Conventional Pretreatments

Conventional pretreatments are the most advanced; they are reliable, closely resemble the notion of unit operation, and the process is well-known and simple to execute [31, 32]. Acid/base, water, and steam are common pretreatment procedures that have been comprehensively discussed in the scientific literature [33]. These methods are the most advanced, and hydrothermal pretreatments (such as liquid hot water and steam explosion) form the foundation of the majority of demo/commercial plant processes [34, 35]. The PROESA® technique (from Versalis) uses a steam-explosion-like pretreatment and is now being deployed in a commercial-scale set up in Crescentino, Italy, having processing capacity of 22 kt/y [36]. Clariant's Sunliquid® technology, which proposes a chemical-free hydrothermal preparation, has also reached commercial scale [37]. Traditional pretreatments aim to maximize sugar conversion, which frequently necessitates the use of severe conditions [38]. The destruction of carbohydrates in the feedstock and the development of inhibitory chemicals for the digestion and fermentation stages are both disadvantages of utilizing high temperatures and/or harsh chemical agents. Other limitations of conventional pretreatments include their significant cost and contribute to the overall cost of the process (up to 30%) [39] and the challenge of chemical catalyst recycling and disposal if they are utilized. Despite the passage of time since their first use, traditional approaches continue to pique researchers' interest, whether in conjunction with other therapies [40, 41] or as part of a biorefinery plan.

3.4.2 Advanced Pretreatments

When the emphasis on pretreatment shifted from the cellulosic fraction to the entire biomass, new technologies has emerged, integrating with the biorefinery concept. All biomass components are significant in this new technique, and not only the amount but also the quality of the goods retrieved is considered. Traditional pretreatments have limitations, and advanced pretreatments are meant to overcome those constraints. They use novel solvents (such as supercritical fluids [42], ionic liquids

(ILs) [43], deep eutectic solvents (DESs) [44], inorganic salts [45]), novel methods (such as popping [46], gamma rays [47], electron beam irradiations [48], novel process strategies (such as milder operating conditions and/or higher solids content), or a combination of several principles to accomplish this. This type of pretreatment has one or more advanced characteristics, like operating at lower temperatures, not creating inhibitors, using fewer chemicals, and having a high selectivity for lignin or effective fractionation [49]. However, as shown in the previously referred studies, most of these revolutionary technologies have only been evaluated in the laboratory or on a semi-pilot size, and hence there is insufficient data to make a reliable techno-economic assessment [50]. Some sophisticated pretreatment methods are promising and have potential for further development; for instance, ILs and DESs are topics of intensive study, with substantial progress made in terms of sustainability, security, price, and performance [49]. Many other approaches, on the other hand, fail to break through due to health or sustainability problems, complex apparatus, low scalability, or poor performance, and research interest fades over time.

3.5 Recent Studies on Different Pretreatment Methods

3.5.1 Physical Pretreatment

Physical pretreatment method can enhance available surface area, pore size, lower cellulose crystallinity and grade of polymerization, reduce the amount of chemicals utilized during method, and reduce chemical waste after treatment [51]. The majority of research has used mechanical methods to lower the particle sizes of biomass feedstocks, such as gridding and chipping. Milling, two-roll milling, ball milling, hammer milling, vibro-energy milling, and colloid milling, are some of the physical processes that can be used as a pretreatment. Irradiation techniques like gamma rays, microwaves, electron beams, and ultrasounds have also been used to enhance the enzymatic degradation or biodegradability of lignocellulosic feedstocks [52–54]. Physical processes, in general, contribute to higher operational costs because of their high energy utilization. Furthermore, compared to chemical pretreatment, physical procedures remove lesser lignin from the cell wall composition. To improve performance, physical treatments are frequently combined with other types of pretreatment.

3.5.2 Chemical Pretreatment

Chemical pretreatment has long been the most common process for saccharification of cellulose. Chemical pretreatments employ acids, alkalis, organic solvents, ozone, and peroxide, among other chemical agents. Concentrated acids, like sulfuric acid and hydrochloric acid, have also been used in the chemical pretreatment, resulting in better enzymatic hydrolysis of fermentable sugars. Chemical pretreatment of a wide range of biomass feedstocks has been recognized to be effective due to the acidic environment's capacity to degrade hemicellulose and lignin to improve cellulose hydrolysis. However, because these potent compounds are toxic and corrosive, they necessitate specialized reactors, which raises the expense [55]. Because of their ability to achieve significant rate of a reaction and improve the subsequent process of biomass degradation, diluted acids are the favored alternative for lignocellulose pretreatment. Furthermore, diluted acids mitigate the drawbacks of concentrated acids, like equipment degradation and the development of inhibitors [56]. For batch processing, dilute acid pretreatment is commonly done at lower temperatures, i.e., temp. lower than 160 °C and high temperatures, i.e., temp. higher than160 °C for continuous-flow processing [57]. The most often utilized acid for pretreatment of diverse feedstocks is dilute sulfuric acid. H2SO4 is the favored acidic pretreatment solution among acidic solutions [58, 59]. In spite the effectiveness of dilute acid, there is still concern regarding the production of inhibitor compounds, for example, hydroxymethylofurfural, acetic acid, and furfural [60]. Neutralization (pH maintenance at 7) step of the pretreatment liquor is also necessary for the subsequent enzymatic saccharification. The procedure of alkaline pretreatment involves adding diluted bases to biomass mixtures, like sodium, potassium hydroxides or anhydrous, calcium, ammonia, etc. to cleave the linkages, present between the lignin and other compounds. The processes speed up the delignification step by breaking the structure of lignin and eliminating the lignin, which reduces the plant cell's mechanical strength. As a result, enzymes have better access to cellulose components of the cell, and sugar liberated from processed material during digestion is better [61]. The alkali pretreatment can be done at room temperature or at high temperatures by soaking the biomass [62]. The cost-effectiveness of the chemicals, the reaction conditions, higher elimination of lignin, and greater possibility of biomass fractionation are the main advantages of alkaline pretreatment. The lignin content of lignocellulosic biomass has been reduced using hydrogen peroxide [63]. The feedstock is saturated in pH-adjusted water containing H_2O_2 at room temperature for 6–24 h in this procedure. Because of good delignification, this method can improve enzymatic hydrolysis results. After basic peroxide usage, bamboo biomass can be transformed into usable sugars with a large yield via enzymatic degradation [64]. In a separate investigation, this method with 7.5 wt% H_2O_2 at 35 °C for 24 h can likewise result in a high conversion up to 96% of rice hulls into hexoses and pentoses after enzymatic degradation. The procedure yielded quantifiable furfural and HMF, making it more accessible for fermentation than a dilute-acid method [65]. The primary disadvantage of the alkali pretreatment is the lengthy time of processing and difficulty neutralizing the post-treatment solution [56]. As a result, many people have tried combining both acid and alkali pretreatments to improve the yield of hydrolysis [66]. As an initial step, acid pretreatment improves removal of hemicellulose. Then, as a second step, alkali pretreatment is used to remove even more lignin, resulting in pure cellulose.

The method of 'ozonolysis' has been successfully used in the processing of lignocellulosic substrates, and it successfully destroys lignin and some proportion of hemicellulose [67]. To reduce the development of inhibitory chemicals, this pretreatment is usually done at lower temperatures [68]. However, because ozonolysis requires a considerable amount of ozone, it can be a costly procedure [69]. Pretreatment with Organosolv can also help to speed up the hydrolysis of lignocelluloses. A delignification agent is an organic (aqueous) solvent [70]. The lignocellulosic feedstocks are mixed with organic liquid and H_2O during the organosolv pretreatment. After that, they are heated to eliminate lignin, degrade hemicellulose, and leave only reactive cellulose in the solid phase [71]. At temperatures ranging from 140 to 200 °C, a variety of organic or aqueous organic solvents, as well as catalysts like acetylsalicylic, salicylic, and oxalic acids, can be used with or without catalysts. Alcohols, esters, ketones, glycols, organic acids, phenols, and other organic solvents have all been used successfully in the past [72]. However, various factors should be considered in order to reduce the process's operational expenses, for instance, the cost of solvents and the ease of solvent recovery. Evaporation and condensation, for example, should be employed to recover the applied solvents, which should then be recycled and reused in following pretreatment batches. Furthermore, solvent removal from pretreated biomass is critical because solvents can hinder enzyme hydrolysis as well as hydrolysate fermentation or digestion [69].

Ionic liquids, like 1-ethyl-3-methylimidazolium acetate and 1-allyl-3-methylimidazolium chlorides, have recently been used to hydrolyze lignocellulosic feedstock components and increase enzymatic degradation in the pretreatment process. Ionic liquids have variable properties depending on the anions and cations used [73]. Small inorganic anions and large organic cations combine to produce these compounds, which are salts. Ionic liquids are typically thought of as green solvents because of their low vapor pressure, wide liquid range, and high dissolving power [74]. Numerous ionic liquids have recently been proven to be capable of hydrolyzing cellulose and lignin, and they can be easily restored from these solutions [75]. However, the optimal dissolution conditions must be identified to avoid cellulose depolymerization and the creation of low molecular products [74].

3.5.3 Physical–Chemical Methods

Physicochemical processes are those that use a combination of physical equipment and chemical chemicals to perform preparation. These procedures can be far more effective than only using physical mechanisms. The most studied approach of physicochemical method is steam explosion. Pressure is abruptly lowered in a steam explosion, causing the materials to decompress explosively [76]. This approach has used higher pressure and higher temperature for a short span of time (pressure varies from 7 to 4.8 MPa and temperature varies from 160 to 260 °C) for a few seconds (e.g., 30 s) to some minutes (e.g., 20 min) [77, 78]. This treatment has a number of advantages, including a quick treatment duration, the absence of chemicals, and minimal energy use, all of which make it economically viable. However, there are concerns concerning lignin removal, xylan breakdown into hemicellulose, and the risk of producing inhibitory compounds during high-temperature processing [56]. Thus, steam explosion pretreatment can be done with the accumulation of various chemicals, like organosolvents and H_2SO_4, to encourage hemicellulose degradation, improve lignin solubilization, and reduce inhibitor synthesis if reduced temperatures are used, resulting in improved enzyme accessibility to cellulose in subsequent processing [79, 80].

AFEX (ammonia fiber/freeze explosion) is a physicochemical pretreatment method [81]. During the AFEX, the biomass is treated under pressures ranging from 0.7 to 2.7 MPa and liquid ammonia at comparatively higher range of temperatures, i.e., 90–100 °C for 30 min before being immediately reduced in pressure. Ammonia loading, water loading, temperature, blowdown pressure, time, and number of treatments are all important parameters in the AFEX process [82]. Ammonia can induce hemicellulose to degrade into oligosaccharides, change the structure of lignin, and cause the biomass cell wall to swell, increasing the accessible surface area for enzyme accessibility [60]. However, when processing biomass with a high lignin concentration, AFEX has a low efficiency and necessitates a huge quantity of ammonia, which requires a high energy input for recycling and recovering [83].

Another pretreatment approach based on the idea of a supercritical carbon dioxide explosion is carbon dioxide explosion [84]. At a lower temperature than steam explosion, this operation is accomplished. A supercritical fluid is a gaseous fluid that is compressed to a liquid-like density at temperatures above its critical point. It is thought that when CO_2 is dissolved in water, it generates carbonic acid, which speeds up the hydrolysis process. Various biomass feedstocks, such as green coconut fiber [85] and soy sauce waste, have been studied for their enzymatic digestibility [86]. Furthermore, after the pretreatment of guayule biomass with supercritical carbon dioxide in the presence of H_2O, modest quantities of the fermentation inhibitory chemicals hydroxymethylfurfural, furfural, and acetic acid were found [87]. The short time required and enhanced sugar output are two major advantages of the supercritical CO_2 process. Though, the supercritical carbon dioxide method may be too costly for commercial use, necessitating additional optimization [88].

The use of liquid hot water (LHW) in the preparation of lignocellulosic materials for bioethanol production is a historical hydrothermal pretreatment method [89]. The pressure is employed in this method to keep water in a liquid form at a high temperature. At temperatures ranging from 160 to 230 °C, pressures greater than 5 MPa are required to keep water in a liquid form [90, 91]. High-pressure water can enter the biomass, hydrate cellulose, and remove hemicellulose and some lignin. This technique dissolves around 40–60% of total biomass, as well as 4–22% of cellulose, 35–60% of lignin, and all of the hemicellulose [92]. No chemicals are used in liquid hot water pretreatment, and the pretreatment parameters are kept simple, relying just on water, heat, and pressure to change the feedstock material. Pretreatment with liquid hot H_2O causes enzyme inhibitors to be released, as well as expose and change lignin that adsorbs enzyme proteins, interfering with enzyme activity delivery to the cellulose substrate [89].

Microwave-assisted pretreatment has recently attracted a lot of attention in the biomass conversion business. Microwave energy is used for both thermal and non-thermal effects in this technology [61]. Microwave-assisted pretreatment can work in tandem with substances like acid, alkali, and organic solvents to boost the yield of the enzymatic hydrolysis that follows [93]. Microwave heating can be used to create a green and low-energy pretreatment procedure. The goals of green extraction are met by the slight use of energy and chemical auxiliaries, like microwave heating for a short period and extremely diluted solvents. Microwave heating technology, on the other hand, has not totally supplanted traditional industrial heating systems. There have been numerous microwave-assisted application studies. The difficulties in employing microwaves to process various biomass waste materials are due to problems with the microwaves themselves as well as the properties of the raw materials. For example, not all materials (such as transparent materials) can be easily heated with microwaves. As a result, the effectiveness of microwave-assisted pretreatment is largely determined by the operating circumstances. Exposure period, solvent type, microwave power, and solid loading (solvent to feed ratio) are the most important elements in sugar recovery during microwave-assisted pretreatments [92]. The use of ultrasonic irradiation to help in the pretreatment process has grown in popularity over the years. It was found to be an effective method for inducing structural changes in lignocellulosic biomass to improve enzymatic saccharification [94]. According to a study, ultrasound-assisted ammonia pretreatment of sugarcane bagasse resulted in a cellulose recovery efficiency of 95.8% and a lignin removal efficiency of 58.1% [95]. Similarly, processing of wasted coffee trash with ultrasound and potassium permanganate resulted in 98% cellulose recovery and 46% delignification [96]. Additionally, two ultrasound-assisted pretreatment techniques, ultrasound-assisted ionic liquid tetrabutylammonium hydroxide ([TBA][OH]) and ultrasound-assisted alkaline, have been successfully used to improve enzymatic saccharification of Eucalyptus [97]. The concerns linked to energy efficiency and the price of the technology are two of the challenges that ultrasound faces as a pretreatment technique. This puts the pretreatment technology's energy efficiency to the test, and it may diminish the technology's cost-effectiveness as it scales up [98].

To improve enzymatic digestibility, mechanical processing via intensive milling combined with metallic salts can be used to pretreat cellulosic materials. The synergistic interplay of mechanical processing and metallic salt in the solid-phase state is used in this combined technique [99]. Intense milling can cause lignocellulosic biomass to shrink in size and become structurally disordered, as well as distort and break chemical bonds. Metal salts are particularly interesting as pretreatment agents since they are less corrosive than inorganic acids [100]. For this preparation, common metal salts like nitrates, chlorides, and sulfates of Mg, Al, Fe, and K are frequently used [101]. Because of weak interaction between metal salts and cellulosic fraction in the solid-phase condition, pretreatment of metal salt is usually done in a liquid state or in an aqueous solution [102]. Metal salt mobility and interaction with LCB components in the solid phase are thus facilitated by mechanical activation by vigorous milling. Furthermore, metal salts can be recycled and reused in the biomass preparation process. Metal salts can be recovered by ultrafiltration in the form of metal hydroxides, which can then be converted back to metal chlorides using conjugated acids. However, there are certain advantages and disadvantages to using inorganic salts. A study found that pretreatment with aluminum chloride, iron chloride, aluminum nitrate, and iron nitrate enhanced the enzymatic hydrolysis of sugarcane by bagasse 79.7%, 65.4%, 65.2%, and 69.0%, respectively. On the other hand, researchers found a wide range of enzymatic hydrolysis yields, ranging from 36.6 to 98.0% [103]. This could be due to the inhibition of metallic ions that remain in the solid portion, which could have negative consequences for enzymes during the bioconversion process [104]. Heavy metal elements, like chromium, have been observed to cause cellulase enzyme inactivation or denaturing. Aside from a lack of knowledge regarding the effects of different inorganic salts on cellulase hydrolysis, there is worry about the economic viability of this type of pretreatment due to the high energy consumption needed by these processes [102].

3.5.4 Biological Methods

Biological methods with microorganisms can be used to process biomass and improve enzymatic degradation or fermentation rate. Many bacteria strains, like *Actinomycetes, Candida, Bacillus,* and *Streptomyces,* as well as certain well-known fungal species, for example—*Ceriporia lancerata, Aspergillus,* and *Cyathus stercolerus,* are used to breakdown LCB [105]. A successful delignification procedure of diverse lignocellulosic feedstocks was done by employing several types of microorganisms that improve the digestibility of organic matter, according to previously published publications [106]. Biological pretreatments need less energy and are carried out in a controlled setting. However, almost all of these techniques are sluggish and need extended incubation times, limiting their industrial use [107] (Table 3.2).

3.6 Estimation of the Cost of Several Pretreatment Processes for a Biorefinery Based on Lignocellulosic Feedstock

Studying the costs of the various components of the entire process can be used to estimate the cost of any established process. These include the cost of the feedstock, apparatus, chemical requirement and cost, and high energy consumption at the time of the process in the case of biomass pretreatment. Various scientific communities have devised different pretreatment procedures at the lab-scale and used mass balance analysis to estimate the overall yield. Flores-Gómez et al. [108] analyzed the overall process of converting agave wastes to ethanol using mass balance analysis. They found that 1000 kg of *A. tequilana* leaf yields 198.4 kg ethanol, as well as 21.6 kg

Table 3.2 Comparative account of different pretreatment methods

Pretreatment	Process type	Advantages	Disadvantages
Physical method	Ball milling, hammer milling	Reduction of particle size and decrystallization of cellulose	High energy input is required
Chemical method	Acid	Hydrolysis of hemicellulose into monosugars and helps in removal of lignin	Higher cost leads to corrosion in equipment and liberation of toxic substances
	Alkaline	Removal of a large amount of lignin and the expansion of the biomass, the available surface area for enzymes increases	Irrecoverable salts develop and are absorbed into biomass, necessitating a long retention time
	Ionic liquids	Mild operational conditions are required for delignification	Higher cost of solvent and need to recover and recycle solvent
	Ozone treatment	Effectively reduce the content of lignin and also inhibitors are not generated	High expanses are needed as huge amount of ozone is required
Physical chemical	Liquid hot water	Hemicellulose is dissolved up to a maximum extent	High temperature and pressure
	Steam explosion	Generate higher fraction of hemicellulose, aids in lignin hydrolysis and is cost-effective	Does not work with softwoods, it also generates inhibitory compounds
	CO_2 Explosion	Does not cause synthesis of inhibitory compounds	Does not modify the structure of lignin and hemicellulose also is expansive process too
	Ammonia fiber explosion (AFEX)	Eliminates lignin and hemicellulose up to some extent and also increases accessible surface area	Recovery of ammonia is required, not suitable for materials having higher content of lignin
Biological	Fungi, Actinomycetes	Requires lesser or no energy input and chemicals, Environment friendly	Slow, i.e., low rate of bioconversion

residual xylose, 14.2 kg of glucose, 22.0 kg of xylo-oligomers, and 29.5 kg galacto-oligomers. The whole method entails pretreatment with AFEXTM, followed by 72 h of simultaneous hydrolysis and fermentation with a mixture of commercial enzymes CTec3 (6.2 kg protein), Htec3 (1.5 kg protein), and yeast. Similarly, another study revealed that the iron chloride-mediated pretreatment of dry rapeseed straw results in mass balance [109]. According to their findings, enzymatic degradation of the solid fraction produced following pretreatment of 100 g of substrate yielded 18.4 g and 1.1 g of glucose and xylose respectively. Mass balance can be thought of as an initial step in determining the viability of a laboratory-developed process being scaled up to pilot scale and then tested on a commercial scale. Kazi et al. [110] conducted a techno-economic study of biochemical ethanol production techniques using several technologies. They discussed the four main pretreatment procedures used to convert maize stover to gasoline (dilute acid, two-stage dilute acid, AFEX, and hot water pretreatment). The dilute acid approach produced the maximum ethanol production of 289 L/t at the lowest fuel cost of 1.36 $/L of gasoline-equivalent. They also advised that the dilute acid pretreatment's operating cost be expressed as a percentage of the entire operating cost. In terms of feedstock, variable operational cost, fixed operational cost, capital depreciation, and average yearly tax, they showed 35.3%, 37.8%, 6.0%, 10.0%, and 10.9%, respectively. The use of waste substrate to generate power makes the overall process more cost-effective and allows for the collection of a 7.1% electricity credit.

Based on the findings, we can conclude that the cost valuation of the total process at the laboratory size is restricted to the raw material utilized, the procedure employed, and the energy required. However, when estimating different financial qualities at a pilot scale or on a big scale, the diverse financial attributes are also significant. As a result, a low-cost pretreatment approach is a key quality in lignocellulosic feedstock-based biorefineries that controls the overall process. The bottleneck of pretreatment technologies, which limits their deployment on a broad scale, will be discussed in the next section. During the initial time of fossil fuel use, humans used the abundant resources without considering the social consequences, such as the environment and human health. Due to increased environmental degradation and global emission over the last four decades, the public and scientific community have become more conscious of the harmful effects of excessive fossil fuel use. As a result, comparable errors cannot be committed in the development of a lignocellulosic biomass-based biorefinery. As a result, in addition to a techno-economic evaluation, the socioeconomic impression of the procedure must be investigated. The possibility for farmers, employment opportunities, environmental impact because of pollutant release, health impact on workers and individuals living nearby, economic growth, and energy safety are some of the socioeconomic indicators. The socioeconomic evaluation aids in the analysis of the social impact of a biorefinery's construction in a region [111].

3.7 Pretreatment Technology Bottlenecks and Future Prospects

The obstacles related to a biorefinery based on lignocellulosic feedstock can be stated as follows:

1. selection and year-round availability of suitable feedstock
2. A productive pretreatment procedure
3. Process costs, such as equipment, reactors, and chemicals as well as process control
4. Toxic chemicals produced during various pretreatment procedures methods
5. Post-pretreatment procedures, such as substrate washing
6. Waste generated throughout the procedure, as well as the environmental implications hazards
7. The chemical or catalysts employed in the process are recycled.

As a result, when selecting a pretreatment strategy, it must meet the following criteria:

- Capable of hydrolyzing 3-D LCB structures, this aids to improve the surface area and porosity of the lignocellulosic feedstock, allowing for better enzyme accessibility and, as a result, higher enzyme performance. This will help to improve hydrolysis.
- Producing cellulose and hemicellulose-rich pretreatment pulp.
- Higher sugar yields, i.e., up to 90%, can be achieved by increasing conversion yields maybe even more.
- Preventing or reducing the development of fermentation inhibitory chemicals like furfural, hydroxymethyl furfurals, acetic acid, and other phenolic or aromatic compounds.
- Increasing lignin recovery for the production of value-added products.
- The costs and stages associated with post-pretreatment activities including washing, pulp separation, and neutralization must be simple or, if possible, eliminated.
- The reactor design should be simple, small, low-cost, and have a high solid loading capacity.
- Less chemical and water requirements; if not, the process should be centered on the recovery and reuse of water and chemicals. The total process will be considerably more sustainable as a result of this.

References

1. Rumyantseva, A., Zhutyaeva, S., & Lazareva, N. (2019). Promotion of investment in renewable energy projects. In *E3S Web of Conferences* (Vol. 91, p. 03006). EDP Sciences.
2. Sahoo, B. K., De, S., & Meikap, B. C. (2011). Improvement of grinding characteristics of Indian coal by microwave pre-treatment. *Fuel Processing Technology, 92*(10), 1920–1928.
3. Pinchuk, V. A., Sharabura, T. A., & Kuzmin, A. V. (2017). Improvement of coal-water fuel combustion characteristics by using of electromagnetic treatment. *Fuel Processing Technology, 167*, 61–68.
4. Melikoglu, M. (2017). Vision 2023: Status quo and future of biomass and coal for sustainable energy generation in Turkey. *Renewable and Sustainable Energy Reviews, 74*, 800–808.
5. Vickers, N. J. (2017). Animal communication: When i'm calling you, will you answer too? *Current Biology, 27*(14), R713–R715.
6. Asomaning, J., Haupt, S., Chae, M., & Bressler, D. C. (2018). Recent developments in microwave-assisted thermal conversion of biomass for fuels and chemicals. *Renewable and Sustainable Energy Reviews, 92*, 642–657.
7. Hassan, S. S., Williams, G. A., & Jaiswal, A. K. (2018). Emerging technologies for the pretreatment of lignocellulosic biomass. *Bioresource Technology, 262*, 310–318.
8. Mohan, S. V., Modestra, J. A., Amulya, K., Butti, S. K., & Velvizhi, G. (2016). A circular bioeconomy with biobased products from CO_2 sequestration. *Trends in Biotechnology, 34*(6), 506–519.
9. Chen, B., & Liu, J. (2004). Properties of lightweight expanded polystyrene concrete reinforced with steel fiber. *Cement and Concrete Research, 34*(7), 1259–1263.
10. Limayem, A., & Andricke, S. (2012). Biomassa lignocelulósica para produção de bioetanol: perspectivas atuais, questões potenciais e perspectivas futuras. *Progresso em Energia e Combustão da Ciência, 38*, 449–467.
11. Vinatier, C., Mrugala, D., Jorgensen, C., Guicheux, J., & Noël, D. (2009). Cartilage engineering: A crucial combination of cells, biomaterials and biofactors. *Trends inBbiotechnology, 27*(5), 307–314.
12. Rabemanolontsoa, H., & Saka, S. (2013). Comparative study on chemical composition of various biomass species. *RSC Advances, 3*(12), 3946–3956.
13. Yu, J., Paterson, N., Blamey, J., & Millan, M. (2017). Cellulose, xylan and lignin interactions during pyrolysis of lignocellulosic biomass. *Fuel, 191*, 140–149.
14. Sun, S., Sun, S., Cao, X., & Sun, R. (2016). The role of pretreatment in improving the enzymatic hydrolysis of lignocellulosic materials. *Bioresource Technology, 199*, 49–58.
15. Werpy, T., & Petersen, G. (2004). Top value-added chemicals from biomass: volume I--results of screening for potential candidates from sugars and synthesis gas (No. DOE/GO-102004–1992). National Renewable Energy Lab., Golden, CO (US).
16. Kassaye, S., Pant, K. K., & Jain, S. (2016). Synergistic effect of ionic liquid and dilute sulphuric acid in the hydrolysis of microcrystalline cellulose. *Fuel Processing Technology, 148*, 289–294.
17. Dora, S., Bhaskar, T., Singh, R., Naik, D. V., & Adhikari, D. K. (2012). Effective catalytic conversion of cellulose into high yields of methyl glucosides over sulfonated carbon based catalyst. *Bioresource Technology, 120*, 318–321.
18. Veluchamy, C., Kalamdhad, A. S., & Gilroyed, B. H. (2018). Advanced pretreatment strategies for bioenergy production from biomass and biowaste. In *Handbook of environmental materials management* (pp.1–19).
19. Quereshi, S., Ahmad, E., Pant, K. K., & Dutta, S. (2019). Insights into microwave-assisted synthesis of 5-ethoxymethylfurfural and ethyl levulinate using tungsten disulfide as a catalyst. *ACS Sustainable Chemistry & Engineering, 8*(4), 1721–1729.
20. Sasaki, M., Adschiri, T., & Arai, K. (2003). Fractionation of sugarcane bagasse by hydrothermal treatment. *Bioresource Technology, 86*(3), 301–304.

21. Ruan, T., Zeng, R., Yin, X. Y., Zhang, S. X., & Yang, Z. H. (2016). Water hyacinth (Eichhornia crassipes) biomass as a biofuel feedstock by enzymatic hydrolysis. *BioResources, 11*(1), 2372–2380.
22. Imman, S., Arnthong, J., Burapatana, V., Champreda, V., & Laosiripojana, N. (2015). Fractionation of rice straw by a single-step solvothermal process: Effects of solvents, acid promoters, and microwave treatment. *Renewable Energy, 83*, 663–673.
23. Christopher, M., Mathew, A. K., Kumar, M. K., Pandey, A., & Sukumaran, R. K. (2017). A biorefinery-based approach for the production of ethanol from enzymatically hydrolysed cotton stalks. *Bioresource Technology, 242*, 178–183.
24. Demirbaş, A. (2005). Thermochemical conversion of biomass to liquid products in the aqueous medium. *Energy Sources, 27*(13), 1235–1243.
25. Ma, Y., Tan, W., Wang, J., Xu, J., Wang, K., & Jiang, J. (2020). Liquefaction of bamboo biomass and production of three fractions containing aromatic compounds. *Journal of Bioresources and Bioproducts, 5*(2), 114–123.
26. Sindhu, R., Kuttiraja, M., Binod, P., Sukumaran, R. K., & Pandey, A. (2014). Physicochemical characterization of alkali pretreated sugarcane tops and optimization of enzymatic saccharification using response surface methodology. *Renewable Energy, 62*, 362–368.
27. Elander, R. T., Dale, B. E., Holtzapple, M., Ladisch, M. R., Lee, Y. Y., Mitchinson, C., Saddler, J. N., & Wyman, C. E. (2009). Summary of findings from the Biomass Refining Consortium for Applied Fundamentals and Innovation (CAFI): Corn stover pretreatment. *Cellulose, 16*(4), 649–659.
28. Alvira, P., Tomás-Pejó, E., Ballesteros, M., & Negro, M. J. (2010). Pretreatment technologies for an efficient bioethanol production process based on enzymatic hydrolysis: A review. *Bioresource Technology, 101*(13), 4851–4861.
29. Galbe, M., & Zacchi, G. (2007). Pretreatment of lignocellulosic materials for efficient bioethanol production. *Biofuels*, 41–65.
30. Galbe, M., & Wallberg, O. (2019). Pretreatment for biorefineries: A review of common methods for efficient utilisation of lignocellulosic materials. *Biotechnology for Biofuels, 12*(1), 1–26.
31. Ruiz, H. A., Conrad, M., Sun, S. N., Sanchez, A., Rocha, G. J., Romaní, A., Castro, E., Torres, A., Rodríguez-Jasso, R. M., Andrade, L. P., & Smirnova, I. (2020). Engineering aspects of hydrothermal pretreatment: From batch to continuous operation, scale-up and pilot reactor under biorefinery concept. *Bioresource Technology, 299*, 122685.
32. Perez-Cantu, L., Schreiber, A., Schütt, F., Saake, B., Kirsch, C., & Smirnova, I. (2013). Comparison of pretreatment methods for rye straw in the second generation biorefinery: Effect on cellulose, hemicellulose and lignin recovery. *Bioresource Technology, 142*, 428–435.
33. Huang, R., Su, R., Qi, W., & He, Z. (2011). Bioconversion of lignocellulose into bioethanol: Process intensification and mechanism research. *Bioenergy Research, 4*(4), 225–245.
34. Alberts, G., Ayuso, M., Bauen, A., Boshell, F., Chudziak, C., Gebauer, J. P., German, L., Kaltschmitt, M., Nattrass, L., Ripken, R., & Robson, P. (2016). Innovation outlook: Advanced liquid biofuels.
35. Kumar, B., Bhardwaj, N., Agrawal, K., Chaturvedi, V., & Verma, P. (2020). Current perspective on pretreatment technologies using lignocellulosic biomass: An emerging biorefinery concept. *Fuel Processing Technology, 199*, 106244.
36. Duque, A., Álvarez, C., Doménech, P., Manzanares, P., & Moreno, A. D. (2021). Advanced bioethanol production: From novel raw materials to integrated biorefineries. *Processes, 9*(2), 206.
37. Wietschel, L., Messmann, L., Thorenz, A., & Tuma, A. (2021). Environmental benefits of large-scale second-generation bioethanol production in the EU: An integrated supply chain network optimization and life cycle assessment approach. *Journal of Industrial Ecology, 25*(3), 677–692.
38. Yoo, C. G., Meng, X., Pu, Y., & Ragauskas, A. J. (2020). The critical role of lignin in lignocellulosic biomass conversion and recent pretreatment strategies: A comprehensive review. *Bioresource Technology, 301*, 122784.

39. Naresh Kumar, M., Ravikumar, R., Thenmozhi, S., Ranjith Kumar, M., & Kirupa Shankar, M. (2019). Choice of pretreatment technology for sustainable production of bioethanol from lignocellulosic biomass: Bottle necks and recommendations. *Waste and Biomass Valorization, 10*(6), 1693–1709.
40. Oliva, J. M., Negro, M. J., Manzanares, P., Ballesteros, I., Chamorro, M. Á., Sáez, F., Ballesteros, M., & Moreno, A. D. (2017). A sequential steam explosion and reactive extrusion pretreatment for lignocellulosic biomass conversion within a fermentation-based biorefinery perspective. *Fermentation, 3*(2), 15.
41. Kim, S. M., Dien, B. S., Tumbleson, M. E., Rausch, K. D., & Singh, V. (2016). Improvement of sugar yields from corn stover using sequential hot water pretreatment and disk milling. *Bioresource Technology, 216*, 706–713.
42. de Carvalho Silvello, M. A., Martínez, J., & Goldbeck, R. (2020). Application of supercritical CO2 treatment enhances enzymatic hydrolysis of sugarcane bagasse. *BioEnergy Research, 13*(3), 786–796.
43. Diez, V., DeWeese, A., Kalb, R. S., Blauch, D. N., & Socha, A. M. (2019). Cellulose dissolution and biomass pretreatment using quaternary ammonium ionic liquids prepared from H-, G-, and S-type lignin-derived benzaldehydes and dimethyl carbonate. *Industrial & Engineering Chemistry Research, 58*(35), 16009–16017.
44. Kim, K. H., Dutta, T., Sun, J., Simmons, B., & Singh, S. (2018). Biomass pretreatment using deep eutectic solvents from lignin derived phenols. *Green Chemistry, 20*(4), 809–815.
45. Kang, K. E., Park, D. H., & Jeong, G. T. (2013). Effects of inorganic salts on pretreatment of Miscanthus straw. *Bioresource Technology, 132*, 160–165.
46. Wi, S. G., Choi, I. S., Kim, K. H., Kim, H. M., & Bae, H. J. (2013). Bioethanol production from rice straw by popping pretreatment. *Biotechnology for Biofuels, 6*(1), 1–7.
47. Liu, Y., Zhou, H., Wang, S., Wang, K., & Su, X. (2015). Comparison of γ-irradiation with other pretreatments followed with simultaneous saccharification and fermentation on bioconversion of microcrystalline cellulose for bioethanol production. *Bioresource Technology, 182*, 289–295.
48. Bak, J. S. (2014). Process evaluation of electron beam irradiation-based biodegradation relevant to lignocellulose bioconversion. *Springerplus, 3*(1), 1–6.
49. Oliva, J. M., Negro, M. J., Álvarez, C., Manzanares, P., & Moreno, A. D. (2020). Fermentation strategies for the efficient use of olive tree pruning biomass from a flexible biorefinery approach. *Fuel, 277*, 118171.
50. Carrozza, C. F., Papa, G., Citterio, A., Sebastiano, R., Simmons, B. A., & Singh, S. (2019). One-pot bio-derived ionic liquid conversion followed by hydrogenolysis reaction for biomass valorization: A promising approach affecting the morphology and quality of lignin of switchgrass and poplar. *Bioresource Technology, 294*, 122214.
51. Moset, V., Xavier, C. D. A. N., Feng, L., Wahid, R., & Møller, H. B. (2018). Combined low thermal alkali addition and mechanical pre-treatment to improve biogas yield from wheat straw. *Journal of Cleaner Production, 172*, 1391–1398.
52. Zhang, H., Zhang, P., Ye, J., Wu, Y., Liu, J., Fang, W., Xu, D., Wang, B., Yan, L., & Zeng, G. (2018). Comparison of various pretreatments for ethanol production enhancement from solid residue after rumen fluid digestion of rice straw. *Bioresource Technology, 247*, 147–156.
53. He, J., & Chen, J. P. (2014). A comprehensive review on biosorption of heavy metals by algal biomass: Materials, performances, chemistry, and modeling simulation tools. *Bioresource Technology, 160*, 67–78.
54. El Achkar, J. H., Lendormi, T., Salameh, D., Louka, N., Maroun, R. G., Lanoisellé, J. L., & Hobaika, Z. (2018). Influence of pretreatment conditions on lignocellulosic fractions and methane production from grape pomace. *Bioresource Technology, 247*, 881–889.
55. Peral, C. (2016). Biomass pretreatment strategies (technologies, environmental performance, economic considerations, industrial implementation). In *Biotransformation of agricultural waste and by-products* (pp. 125–160). Elsevier.
56. Bhutto, A. W., Qureshi, K., Harijan, K., Abro, R., Abbas, T., Bazmi, A. A., Karim, S., & Yu, G. (2017). Insight into progress in pre-treatment of lignocellulosic biomass. *Energy, 122*, 724–745.

57. Cheng, J. J. (2017). Anaerobic digestion for biogas production. In *Biomass to renewable energy processes* (pp. 143–194). CRC Press.
58. Santos, C. C., de Souza, W., Sant'Anna, C., & Brienzo, M. (2018). Elephant grass leaves have lower recalcitrance to acid pretreatment than stems, with higher potential for ethanol production. *Industrial Crops and Products, 111*, 193–200
59. Sarip, H., Hossain, M. S., Azemi, M., & Allaf, K. (2016). A review of the thermal pretreatment of lignocellulosic biomass towards glucose production: Autohydrolysis with DIC technology. *BioResources, 11*(4), 10625–10653.
60. Jędrzejczyk, M., Soszka, E., Czapnik, M., Ruppert, A. M., & Grams, J. (2019). Physical and chemical pretreatment of lignocellulosic biomass. In *Second and third generation of feedstocks* (pp. 143–196). Elsevier.
61. Ethaib, S., Omar, R., Mazlina, M. S., Radiah, A. D., & Zuwaini, M. (2020, June). Evaluation solvent level effect on sugar yield during microwave-assisted pretreatment. In *IOP Conference Series: Materials Science and Engineering* (Vol. 871, No. 1, p. 012034). IOP Publishing.
62. Sivanarutselvi, S., Poornima, P., Muthukumar, K., & Velan, M. (2019). Studies on effect of alkali pretreatment of banana pseudostem for fermentable sugar production for biobutanol production. *Journal of Environmental Biology, 40*(3), 393–399.
63. Ho, M. C., Ong, V. Z., & Wu, T. Y. (2019). Potential use of alkaline hydrogen peroxide in lignocellulosic biomass pretreatment and valorization—A review. *Renewable and Sustainable Energy Reviews, 112*, 75–86.
64. Song, X., Jiang, Y., Rong, X., Wei, W., Wang, S., & Nie, S. (2016). Surface characterization and chemical analysis of bamboo substrates pretreated by alkali hydrogen peroxide. *Bioresource Technology, 216*, 1098–1101.
65. Sahay, S. (2020). Impact of Pretreatment Technologies for Biomass to Biofuel Production. In *Substrate Analysis for Effective Biofuels Production* (pp. 173–216). Springer
66. Weerasai, K., Suriyachai, N., Poonsrisawat, A., Arnthong, J., Unrean, P., Laosiripojana, N., & Champreda, V. (2014). Sequential acid and alkaline pretreatment of rice straw for bioethanol fermentation. *BioResources, 9*(4), 5988–6001.
67. Ab Rasid, N. S., Zainol, M. M., & Amin, N. A. S. (2020). Pretreatment of agroindustry waste by ozonolysis for synthesis of biorefinery products. In *Refining Biomass Residues for Sustainable Energy and Bioproducts* (pp. 303–336). Academic Press.
68. Ballesteros, L. F., Michelin, M., Vicente, A. A., Teixeira, J. A., & Cerqueira, M. Â. (2018). Lignocellulosic materials: sources and processing technologies. In *Lignocellulosic Materials and Their Use in Bio-based Packaging* (pp. 13–33). Springer.
69. Sun, Y., & Cheng, J. (2002). Hydrolysis of lignocellulosic materials for ethanol production: A review. *Bioresource Technology, 83*(1), 1–11.
70. Nitsos, C., Rova, U., & Christakopoulos, P. (2018). Organosolv fractionation of softwood biomass for biofuel and biorefinery applications. *Energies, 11*(1), 50.
71. Choi, J. H., Jang, S. K., Kim, J. H., Park, S. Y., Kim, J. C., Jeong, H., Kim, H. Y., & Choi, I. G. (2019). Simultaneous production of glucose, furfural, and ethanol organosolv lignin for total utilization of high recalcitrant biomass by organosolv pretreatment. *Renewable Energy, 130*, 952–960.
72. Zhang, K., Pei, Z., & Wang, D. (2016). Organic solvent pretreatment of lignocellulosic biomass for biofuels and biochemicals: A review. *Bioresource Technology, 199*, 21–33.
73. Yoo, C. G., Pu, Y., & Ragauskas, A. J. (2017). Ionic liquids: Promising green solvents for lignocellulosic biomass utilization. *Current Opinion in Green and Sustainable Chemistry, 5*, 5–11.
74. Mäki-Arvela, P., Anugwom, I., Virtanen, P., Sjöholm, R., & Mikkola, J. P. (2010). Dissolution of lignocellulosic materials and its constituents using ionic liquids—a review. *Industrial Crops and Products, 32*(3), 175–201.
75. Reddy, P. (2015). A critical review of ionic liquids for the pretreatment of lignocellulosic biomass. *South African Journal of Science, 111*(11–12), 1–9.
76. Wu, Y., Hao, Y., Wei, X., Shen, Q., Ding, X., Wang, L., Zhao, H., & Lu, Y. (2017). Impairment of NADH dehydrogenase and regulation of anaerobic metabolism by the small RNA RyhB

and NadE for improved biohydrogen production in Enterobacter aerogenes. *Biotechnology for biofuels, 10*(1), 1–15.
77. Baêta, B. E. L., de Miranda Cordeiro, P. H., Passos, F., Gurgel, L. V. A., de Aquino, S. F., & Fdz-Polanco, F. (2017). Steam explosion pretreatment improved the biomethanization of coffee husks. *Bioresource Technology, 245*, 66–72.
78. Bonfiglio, F., Cagno, M., Rey, F., Torres, M., Böthig, S., Menéndez, P., & Mussatto, S. I. (2019). Pretreatment of switchgrass by steam explosion in a semi-continuous pre-pilot reactor. *Biomass and Bioenergy, 121*, 41–47.
79. Guerrero, A. B., Ballesteros, I., & Ballesteros, M. (2017). Optimal conditions of acid-catalysed steam explosion pretreatment of banana lignocellulosic biomass for fermentable sugar production. *Journal of Chemical Technology & Biotechnology, 92*(9), 2351–2359.
80. Martino, D. C., Colodette, J. L., Chandra, R., & Saddler, J. (2017). Steam explosion pretreatment used to remove hemicellulose to enhance the production of a eucalyptus organosolv dissolving pulp. *Wood Science and Technology, 51*(3), 557–569.
81. Mathew, A. K., Parameshwaran, B., Sukumaran, R. K., & Pandey, A. (2016). An evaluation of dilute acid and ammonia fiber explosion pretreatment for cellulosic ethanol production. *Bioresource Technology, 199*, 13–20.
82. Chundawat, S. P. S., Pal, R. K., Zhao, C., Campbell, T., Teymouri, F., Videto, J., Nielson, C., Wieferich, B., Sousa, L., and Dale, B. E., *et al.* 2020. "Ammonia fiber expansion (AFEX) pretreatment of lignocellulosic biomass," *Journal of Visualized Experiments* 2020(158), e57488.
83. Rabemanolontsoa, H., & Saka, S. (2016). Various pretreatments of lignocellulosics. *Bioresource Technology, 199*, 83–91.
84. Al Afif, R., Wendland, M., Amon, T., & Pfeifer, C. (2020). Supercritical carbon dioxide enhanced pre-treatment of cotton stalks for methane production. *Energy, 194*, 116903.
85. Putrino, F. M., Tedesco, M., Bodini, R. B., & de Oliveira, A. L. (2020). Study of supercritical carbon dioxide pretreatment processes on green coconut fiber to enhance enzymatic hydrolysis of cellulose. *Bioresource Technology, 309*, 123387.
86. Xiang, C., Liu, S. Y., Fu, Y., & Chang, J. (2019). A quick method for producing biodiesel from soy sauce residue under supercritical carbon dioxide. *Renewable Energy, 134*, 739–744.
87. Islam, S. M., Elliott, J. R., & Ju, L. K. (2018). Minimization of fermentation inhibitor generation by carbon dioxide-water based pretreatment and enzyme hydrolysis of guayule biomass. *Bioresource Technology, 251*, 84–92.
88. Carneiro, T. F., Timko, M., Prado, J. M., & Berni, M. (2016). Biomass pretreatment with carbon dioxide. In *Biomass fractionation technologies for a lignocellulosic feedstock based biorefinery* (pp. 385–407). Elsevier.
89. Ximenes, E., Farinas, C. S., Kim, Y., & Ladisch, M. R. (2017). Hydrothermal pretreatment of lignocellulosic biomass for bioethanol production. In *Hydrothermal processing in biorefineries* (pp. 181–205). Springer.
90. Kim, Y., Kreke, T., Mosier, N. S., & Ladisch, M. R. (2014). Severity factor coefficients for subcritical liquid hot water pretreatment of hardwood chips. *Biotechnology and Bioengineering, 111*(2), 254–263.
91. Ko, J. K., Kim, Y., Ximenes, E., & Ladisch, M. R. (2015). Effect of liquid hot water pretreatment severity on properties of hardwood lignin and enzymatic hydrolysis of cellulose. *Biotechnology and Bioengineering, 112*(2), 252–262.
92. Kumar, S., Kothari, U., Kong, L., Lee, Y. Y., & Gupta, R. B. (2011). Hydrothermal pretreatment of switchgrass and corn stover for production of ethanol and carbon microspheres. *Biomass and Bioenergy, 35*(2), 956–968.
93. Zhang, Y., Wang, Z., Feng, J., & Pan, H. (2020). Maximizing utilization of poplar wood by microwave-assisted pretreatment with methanol/dioxane binary solvent. *Bioresource Technology, 300*, 122657.
94. Wang, S., Li, F., Zhang, P., Jin, S., Tao, X., Tang, X., Ye, J., Nabi, M., & Wang, H. (2017). Ultrasound assisted alkaline pretreatment to enhance enzymatic saccharification of grass clipping. *Energy Conversion and Management, 149*, 409–415.

95. Ramadoss, G., & Muthukumar, K. (2014). Ultrasound assisted ammonia pretreatment of sugarcane bagasse for fermentable sugar production. *Biochemical Engineering Journal, 83*, 33–41.
96. Ravindran, R., Jaiswal, S., Abu-Ghannam, N., & Jaiswal, A. K. (2017). Evaluation of ultrasound assisted potassium permanganate pre-treatment of spent coffee waste. *Bioresource Technology, 224*, 680–687.
97. Wang, Z., Hou, X., Sun, J., Li, M., Chen, Z., & Gao, Z. (2018). Comparison of ultrasound-assisted ionic liquid and alkaline pretreatment of Eucalyptus for enhancing enzymatic saccharification. *Bioresource Technology, 254*, 145–150.
98. Bussemaker, M. J., & Zhang, D. (2013). Effect of ultrasound on lignocellulosic biomass as a pretreatment for biorefinery and biofuel applications. *Industrial & Engineering Chemistry Research, 52*(10), 3563–3580.
99. Zhang, Y., Li, Q., Su, J., Lin, Y., Huang, Z., Lu, Y., Sun, G., Yang, M., Huang, A., Hu, H., & Zhu, Y. (2015). A green and efficient technology for the degradation of cellulosic materials: Structure changes and enhanced enzymatic hydrolysis of natural cellulose pretreated by synergistic interaction of mechanical activation and metal salt. *Bioresource Technology, 177*, 176–181.
100. Xu, J., Xu, J., Zhang, S., Xia, J., Liu, X., Chu, X., Duan, J., & Li, X. (2018). Synergistic effects of metal salt and ionic liquid on the pretreatment of sugarcane bagasse for enhanced enzymatic hydrolysis. *Bioresource Technology, 249*, 1058–1061.
101. George, A., Brandt, A., Tran, K., Zahari, S. M. N. S., Klein-Marcuschamer, D., Sun, N., Sathitsuksanoh, N., Shi, J., Stavila, V., Parthasarathi, R., & Singh, S. (2015). Design of low-cost ionic liquids for lignocellulosic biomass pretreatment. *Green Chemistry, 17*(3), 1728–1734.
102. Loow, Y. L., Wu, T. Y., Tan, K. A., Lim, Y. S., Siow, L. F., Md. Jahim, J., Mohammad, A. W., & Teoh, W. H. (2015). Recent advances in the application of inorganic salt pretreatment for transforming lignocellulosic biomass into reducing sugars. *Journal of Agricultural and Food Chemistry, 63*(38), 8349–8363
103. López-Linares, J. C., Romero, I., Moya, M., Cara, C., Ruiz, E., & Castro, E. (2013). Pretreatment of olive tree biomass with $FeCl_3$ prior enzymatic hydrolysis. *Bioresource Technology, 128*, 180–187.
104. Tejirian, A., & Xu, F. (2010). Inhibition of cellulase-catalyzed lignocellulosic hydrolysis by iron and oxidative metal ions and complexes. *Applied and Environmental Microbiology, 76*(23), 7673–7682.
105. Pandey, A., Tiwari, S., Jadhav, S. K., & Tiwari, K. L. (2014). Efficient microorganism for bioethanol production from lignocellulosic *Azolla. Research Journal of Environmental Sciences, 8*(6), 350–355.
106. Maurya, D. P., Singla, A., & Negi, S. (2015). An overview of key pretreatment processes for biological conversion of lignocellulosic biomass to bioethanol. *3 Biotech, 5*(5), 597–609.
107. Zabed, H. M., Akter, S., Yun, J., Zhang, G., Awad, F. N., Qi, X., & Sahu, J. N. (2019). Recent advances in biological pretreatment of microalgae and lignocellulosic biomass for biofuel production. *Renewable and Sustainable Energy Reviews, 105*, 105–128.
108. Flores-Gómez, C. A., Escamilla Silva, E. M., Zhong, C., Dale, B. E., da Costa Sousa, L., & Balan, V. (2018). Conversion of lignocellulosic agave residues into liquid biofuels using an AFEX™-based biorefinery. *Biotechnology for Biofuels, 11*(1), 1–18.
109. Romero, I., López-Linares, J. C., Moya, M., & Castro, E. (2018). Optimization of sugar recovery from rapeseed straw pretreated with $FeCl_3$. *Bioresource Technology, 268*, 204–211.
110. Kazi, F. K., Fortman, J. A., Anex, R. P., Hsu, D. D., Aden, A., Dutta, A., & Kothandaraman, G. (2010). Techno-economic comparison of process technologies for biochemical ethanol production from corn stover. *Fuel, 89*, S20–S28.
111. Fedorova, E., Caló, A., & Pongrácz, E. (2019). Balancing socio-efficiency and resilience of energy provisioning on a regional level, Case Oulun Energia in Finland. *Clean Technologies, 1*(1), 273–293.

Chapter 4
A Feasible Approach for Bioethanol Production Using Conventional and New Feedstocks

Abstract Current energy demand, rising oil prices, and the consequences of using fossil fuels have increased the desire for alternate energy sources. Bioethanol obtained from new feedstocks has gained attention and led to the production of advanced bioethanol over conventional. In terms of sustainability, the production of so-called advanced bioethanol has various advantages over typical bioethanol production procedures. This could involve the utilization of non-food crops or residual biomass as a raw material, as well as a greater ability to reduce greenhouse gas emissions. The current chapter focuses on recent advances in the production of advanced bioethanol, highlighting current results from novel feedstock sources such as the municipal solid waste and certain industrial waste (e.g., residues from the paper industry, food, and beverage industries) and lignocellulosic feedstocks.

Keywords Bioethanol · Lignocellulose · Nanotechnology · Sustainability · Economy

4.1 Introduction

The usage of fossil fuels and its consequences on climate change are driving forces behind the quest for the renewable sources, which could be a ray of hope for long-term sustainability [1, 2]. Petroleum-based sources are limited, create greenhouse gas emissions, and pollute the air, which has prompted the search for alternative biofuels [3]. Biofuels, on the other hand, are environmentally beneficial and have a huge potential for reducing GHG emissions. Furthermore, unlike water, photovoltaic, and wind energy, these fuels can be conveniently stored as liquid fuels [4, 5]. Bioethanol produced from agro-residues of cereal crops has a lot of potential because of the higher yield of hybrids, the availability of feedstock and know-how method, the circular economic approach to biorefinery development, and the fact that it is both cost-effective and environmentally friendly [6, 7].

Biofuel, also known as bioethanol, is a type of alternative energy resource that has gotten a lot of interest across the world. Bioethanol is currently the sole substitute for gasoline that may be used without requiring any modifications to the way gasoline is distributed. Furthermore, because the carbon dioxide produced during bioethanol

N. K. Aggarwal et al., *Bioethanol Production*, Green Chemistry and Sustainable Technology, https://doi.org/10.1007/978-3-031-05091-6_4

burning is similar to that used by plants in the atmosphere for their growth and metabolism, it does not contribute to greenhouse gas emissions. Bioethanol is used in a variety of businesses around the world. Bioethanol is used as a fuel in combustion operations, as well as in biofuel cells, as a feedstock for chemical businesses, and as a fuel in cogeneration systems. Furthermore, one of bioethanol's many applications is as an alternative fuel to gasoline in motors [8]. For various goals, researchers are investigating various methods to increase the processes and output of bioethanol using perfect and economical feedstock [9, 10].

4.2 Conventional Feedstocks for Production of Bioethanol

4.2.1 Cereal Based Substrates for Synthesis of Bioethanol

Starch and protein are the principal elements of cereal grains like maize, barley, rice, wheat, sorghum, rye, and oats, whereas phytic acid, vitamins, lipids, carbohydrates, and minerals are minor constituents. Cereals' high starch content made them a potential ethanol substrate [11, 12]. To liberate starch from the feedstock, the grains are either dry ground or wet ground as the first stage in the ethanol manufacturing process. This is followed by gelatinization, which involves heating the starch to a high temperature (Table 4.1). During liquefaction and saccharification, amylolytic enzymes act on the viscous slurry acquired by gelatinization to release sugars, which are subsequently acted on by yeast or any other microbe for anaerobic fermentation to produce ethanol. Distillation, rectification, and dehydration operations separate and concentrate the ethanol produced with CO_2. The amount of ethanol produced is determined by the starch amount of the feedstock, process characteristics, and the ethanol production process. Aside from separate hydrolysis and fermentation, simultaneous saccharification and fermentation, and simultaneous saccharification and co-fermentation, an integrated method is being used to boost ethanol output while lowering the cost and time of the process.

Table 4.1 Content of starch and ethanol yield from different cereal feedstocks

Substrate	Starch content (%)	Ethanol yield (L/100 kg)
Rice	55–70	48–57
Sorghum	55–65	36–42
Wheat	58–62	36–39
Maize	60–63	38–40
Rye	56–70	35–42
Oats	54–64	36–42

4.2.2 Cereal Based Substrates Derived from Agro-residues for Synthesis of Ethanol

When the cereals are harvested, residue that remains on the farm is used as animal feed to a limited extent, with the remainder being burned. The smoke produced by burning poses a serious health risk. The grain waste is lignocellulosic in origin, with lignin (10–20%), cellulose (40–50%), and hemicellulose (20–30%) as the main constituents. [13]. Cellulose is a polymeric structure made up of glucose, that gives plants their mechanical strength, whereas hemicellulose is a hexose and pentose heteropolysaccharide. Non-covalent attractions bind cellulose and hemicellulose to one another. Similarly, lignin, which is made up of different alcohols like sinapyl, coumaryl, and coniferyl, forms a protective cover around holocelluloses. The chemical makeup of several cereal wastes has been compiled (Table 4.2) [14, 15].

Thermochemical processes like gasification and pyrolysis are used to manufacture value-added compounds like biofuels and other chemicals derived from lignocellulose. Thermolysis of feedstock produces syngas and bio-crude, which are used as drop-in fuel precursors. Pyrolysis is a process in which biomass is heated to 500–600° Fahrenheit in the absence of oxygen to produce bio-oil, which is then hydroprocessed into a drop-in fuel precursor. Feedstock can be converted to liquid fuel via gasification at greater temperatures, above 700 °C, and under controlled oxygen. Microorganisms such as *Butyribacterium methylotrophicum, Clostridium ljungadahlii, Clostridium autoethanogenum, Clostridium carboxydivorans, Methanosarcina barkeri,* and *Rhodospirillum rubrum* [16] or metal catalysts [Fischer-Tropsch (FT) synthesis] such as aluminum, cobalt, and others can convert the syngas into bioethanol which is produced during this process [17].

Table 4.2 Lignocellulosic compositional analysis of various Agri residual waste-derived from cereals

Substrate	Cellulose (% dry wt)	Hemicellulose (% dry wt)	Lignin (% dry wt)
Wheat straw	33–41	26–32	13–19
Cornstalk	39–47	26–31	3–5
Rice straw	28–36	23–28	12–14
Corn cobs	45	35	15
Sorghum straw	32	24	13
Sweet sorghum bagasse	34–45	18–28	14–22
Barley straw	31–45	27–38	14–19
Corn stover	38–40	28	7–21
Wheat husk	36	18	16
Rye husk	26	16	13
Sorghum stalks	27	25	11

4.3 An Account of New Feedstocks for Production of Bioethanol

One of the most viable feedstocks for sustainable biofuel generation is lignocellulosic biomass. In terms of cost, lignocellulose is easily available in huge amount and can be produced locally at a reasonable cost. Agricultural residues or forest leftovers (such as straws, sugarcane bagasse, corn stover, sawdust, pruning, softwood, hardwood, and bark residues), energy crops, and the organic part of domestic and industrial wastes are all examples of lignocellulosic materials. These materials are plentiful on the planet and have little economic worth in many circumstances. Furthermore, because this sort of substate cannot be employed as a food source, there is no potential for conflict with the food business. As a result, lignocellulosic bioethanol is a potential and environmentally benign alternative to first-generation bioethanol made from food crop waste.

The major components of lignocellulosic biomass are cellulose, hemicellulose, and lignin. -D-glucose units are joined by -1-4-glycosidic linkages to form cellulose, a linear polymer. Because of hydrogen bonding and Van der Waal's forces, cellulose in lignocellulosic biomass is structured into microfibrils by cross-linking between multiple hydroxyl groups. This forms a hard structure and is the most important constituent of plant cell walls. Hemicellulose is a heteropolysaccharide made up of sugars with 50–200 units called pentoses (like xylose and arabinose) and hexoses (like glucose, mannose, and galactose). Hemicellulose is highly branched structure with acetyl groups and modest quantities of glucuronic acid, and their composition varies depending on the biomass. Lignin is the third main component of lignocellulosic biomass, and it gives the plant its structural strength. Lignin is an aromatic or cyclic structure, which is made up of three units, i.e., coniferyl, sinapyl, and p-coumaryl alcohol, that is heterogeneous, amorphous, and extremely branched.

Biotechnological procedures can convert the sugars in these feedstocks into bioethanol in multiple phases and with a variety of configuration options. Biochemical bioethanol production systems have a theoretical maximum yield of 0.51 g of bioethanol per gram of monosaccharide. Different factors, however, influence the efficiency and yield of bioethanol synthesis. One of the most significant aspects to consider during advanced bioethanol production is biomass recalcitrance, which reduces carbohydrate accessibility and hence limits the liberation of fermentable sugars. To address biomass resistance, various pretreatment procedures have been developed to provide carbohydrate accessibility.

4.3.1 Utilization of Energy Crops

Energy crops are those crops whose portion or total yield is utilized as a raw material to obtain valuable energy. These crops, on average, create a lot of biomass per unit of land and time. Rapid growth and the ability to grow in harsh conditions of weather and soil are the most important features to look for when choosing energy crops [18]. Herbaceous and woody energy crops are the two main categories of energy crops. Switchgrass, giant reed, and miscanthus are among the perennial grasses that make up herbaceous energy crops. Poplar and eucalyptus, on the other hand, seem to have been a short rotation woody crops with a rather high growth rate. In poor soils, both woody crops and herbaceous crops may help to prevent soil erosion and improve soil carbon content and fertility. Herbaceous energy crops can also be grown on less fertile land without compromising bioethanol's fundamental properties [19, 20].

The most studied lignocellulosic energy crops are switchgrass and *Miscanthus* spp. which evaluated the conversion of switchgrass after exposing it to various pretreatment processes (acid, alkali), with methanol proving to be the most successful pretreatment process, yielding 0.32 g of ethanol/g of glucose with conversion rates of 97% [21].

4.3.2 Exploitation of Forest Biomass

Forestry feedstock has been used as a substrate for bioethanol synthesis, primarily woody components like branches, leaves, and lops. Softwood (from gymnosperms) and hardwood (from trees) are two types of wood (from angiosperms). Growth rates and densities are two major distinctions between softwoods and hardwoods. Hardwoods grow more slowly than softwoods and are consequently denser [22].

This organization is vital to the lowering of carbon dioxide levels in the atmosphere as well as the preservation of marginal land. In comparison to other raw materials, one of the key benefits of employing forestry biomass as a feedstock is the flexibility in harvesting time, as these materials are not seasonal.

Eucalyptus nitens bark is a forestry resource that is widely obtained through the production of eucalyptus trees in the pulping business. For every 100 tons of pulp product produced, 20 tons of bark can be produced [23]. Roman et al. [24], employing organosolv as a pretreatment method and simultaneous saccharification with fermentation (SSF) as a process approach, utilized this residue as an alternate source for making 252 L of bioethanol per ton of biomass.

Boards manufactured from wood chips, branches, and reeds are more intriguing forest-derived leftovers. After use, these panels are usually burned. These materials, on the other hand, can be used to make bioethanol because they are mostly made up of lignocellulose (which accounts for more than 85% of their makeup). Zhao et al. [25] explored the application of these materials for bioethanol synthesis, achieving bioethanol conversion yields of 84–95% of theoretical utilizing phosphoric acid and H_2O_2 as pretreatment methods followed by 72 h of SSF.

4.3.3 *Utilization of Agricultural Residues*

Agricultural leftovers are described as "a crop lost during the year at all stages between the farm and the home level during processing, storage, and transport," according to the United Nations' Food and Agriculture Organization (FAO). This covers both field and processing residues and includes components like straw, bagasse, stubble, seeds, and husks. By 2030, the number of agricultural residues accessible for bioenergy generation in the United States alone is expected to be 240 million dry t/year [26]. Rice, wheat, corn, and sugarcane account for a large share of agricultural wastes, with rice straw being the most common agricultural residue globally.

4.3.4 *Use of Industrial Waste*

All byproduct streams from established industrial areas like pulp or paper, textiles, food, as well as 1G biodiesel and bioethanol-based activities, are considered industrial wastes. Utilization of this kind of biomass helps to minimize the environmental impact of these commercial/industrial processes by lowering net carbon dioxide emissions, minimizing our reliance on petroleum-based options, and enhancing the economic efficacy of the processes by putting a value on what would otherwise be rejected as waste [27].

Brewers' spent grains account for about 85% of the total amount of byproducts produced by the brewery when it comes to food-derived wastes. This residual waste may have a large quantity of fermentable sugars, ranging from 10 to 25% and 15 to 30% for both cellulose and hemicellulose, respectively, depending on the skill used in each brewery and the kind of crop. As a result, these sugars are an appealing substrate for bioethanol production. Researchers studied the extraction of fermentable sugars from brewer's discarded grains obtained from a commercial brewing industry in Spain by treating the waste with dilute H_2SO_4 under mild circumstances (i.e., 130 °C and 26 min) [28]. This method allowed up to 94% of the initial fermentable sugars in the substrate to be recovered.

Bioethanol and other high-value-added products can be made from waste from the pulp and paper industries. In 2015, 400 million tons of paper and paperboard were produced, as well as 188 million tons of raw pulp [29]. Major residual waste obtained from paper industry is potent for bioethanol production including spent sulfite liquor (SSL) and pulp and paper mill sludge (PPMS). It is worth noting that SSL obtained from hardwood has a larger proportion of pentoses than SSL derived from softwood. For converting hardwood liquor into bioethanol, pentose-fermenting bacteria such as *Scheffersomyces stipitis* (previously *Pichia stipitis*) must be recognized [30]. PPMS is a solid waste that contains a higher content of glucan and can be transformed into a variety of value-added compounds. The type of biomass has a direct impact on the chemical composition of PPMS (organic content, ash, pH, and so on). The papermaking process and wastewater cleansing technology are two more elements that determine the composition of PPMS [31]. Schroeder et al. [32] investigated the synthesis of bioethanol from PPMS using two different process configurations: separate hydrolysis and fermentation (SHF) and separate hydrolysis and fermentation (SSF). The SSF technique produced the maximum ethanol yield (55.7%), which was also achieved in a lower overall process time (72 h for the SSF vs. 84 h for the SHF process). Bioethanol synthesis has also been investigated using materials based on cotton, jute (*Corchorus* sp.) biomass, and mesta (Hibiscus spp.) biomass [33–35]. The utilization of these feedstocks helps to cut down on the amount of waste produced in the textile sector. Jute feedstock has recently been touted as a possible bioethanol feedstock, particularly in case of India, when 1.8 billion kg of this feedstock was generated in year 2018–2019 [36].

4.3.5 Municipal Waste Utilization

In Europe, the average quantity of municipal solid waste (MSW) produced per person per year is 475 kg [37]. Because of exponential growth in population of world, there has been huge consumption of natural sources and energy, and this number is always rising. MSW consists of biodegradable yard waste, domestic waste from homes, offices, restaurants, canteens, wholesale and retail establishments, as well as equivalent waste from food manufacturing companies. The organic fraction of municipal solid waste (OFMSW) is the most significant component of MSW, accounting for approximately 40–50% of the total dry matter. Carbohydrates (30–40%), proteins (5–15%), and lipids (10–15%), make up the majority of OFMSW (dry weight). Furthermore, inert materials like plastic, textiles, and glass are commonly found in OFMSWs, and their concentrations are highly dependent on the collection system utilized. A large volume of inert materials can cause a variety of technological issues and reduce the effectiveness of the valorization process.

The sorting mechanism, seasonality, population, dietary patterns, and socioeconomic situations all influence the features and production rates of OFMSW [38]. Because of its complexity and heterogeneity, OFMW is thought to be a difficult substrate. Moreno et al. [39] compared two biowaste collection sorting systems (source sorted (SSOFMW) and nonsorted (NSOFMW)) in order to employ them as feedstock for both synthesis of bioethanol as well as biogas. After exposing these two substrates to non-isothermal SSF procedures, the maximum ethanol concentrations attained were 51 and 26 g/L for source sorted and nonsorted waste, respectively, illustrating the need of segregating biowaste at the source.

4.4 Genetic Engineering for Enhancing the Production Yield of Bioethanol

Genetic engineering of lignocellulosic biomass is an intriguing method for improving carbohydrate accessibility during conversion processes. Secondary cell walls have a complicated chemical composition and structure that creates a physical barrier that is difficult to breach. Different approaches to genetically altering lignocellulosic biomass and enhancing conversion efficiencies have been proposed to reduce biomass recalcitrance. Furthermore, this technique permits the tuned qualities to be passed along to subsequent generations, making the process more efficient over time. Regardless, the genetic modifications introduced in these organisms may have an impact on cell growth of plant and developmental processes, thereby lowering biomass outputs [40].

Plants can be genetically modified to promote the synthesis of polymer in cell walls (particularly cellulose), minimize cellulose crystallinity, and change the structural complexity of lignin. Lignin Alteration is one of these techniques that has been specifically aimed at lowering biomass recalcitrance. Changes in phenolic metabolism have been used to change the structure of lignin. The p-hydroxphenyl (H), guaiacyl (G), and syringyl (S) units in lignin polymer are derived from the p-coumaryl, coniferyl, and sinapyl alcohols, respectively. The proportion of these units varies by tissue and plant type. Gymnosperm lignins, for example, often have a higher proportion of G units, whereas angiosperm lignins primarily contain both G and S units [41].

Several enzymes are involved in the manufacture of monolignols, making them interesting candidates for genetic engineering [42]. 4-coumarate, CoA ligase (4CL), which catalyzes the bioconversion of p-coumaric acid to p-coumaroyl-CoA, is a crucial enzyme in the preliminary part of this biosynthetic process (intermediate metabolite in the phenylpropanoid pathway). Park et al. [43] created genetically modified switchgrass plants that were knock-out mutant, with an approximately 8–30% drop in lignin concentration by removing Pv4CL1 gene (which encodes the 4CL enzyme). Without acid pretreatment, these mutants improved both glucose as well as xylose synthesis, compared to control tests.

During plant engineering, cellulose content and lowering its crystallinity is an important goal in addition to changing the lignin content. Yang et al. [44] studied the overexpression of gene named PdDUF266A present in Populus in this setting. DUF266 proteins are classed as "not classified glycosyltransferases" (GTs) and may hence play an important role in polysaccharide production. The total cellulose content of Populus increased by 37.1% when the PdDUF266A gene was overexpressed, resulting in a 38% increase in total sugar produced during the consequent saccharification.

In summary, genetic engineering of plant cell walls has a lot of potential for enhancing the efficiency of lignocellulose conversion processes in a bio-based economy, but more research is required to fully grasp the potential of these tactics and mitigate the negative effects of such changes.

4.5 Aquatic Weeds: Novel Substrate for Production of Bioethanol

Aquatic weeds have a substantial cellulose, starch, and lipid content, making them an ideal feedstock for bioenergy generation. Diverse research activities have been done globally in recent years in attempt to manufacture biofuel from a variety of aquatic biomasses. Because of its plentiful availability, extraordinary adaptation capabilities, and tremendous growth rate, *E. crassipes,* i.e., water hyacinth is popular aquatic weed [45]. Unfortunately, water hyacinth growth is difficult to control. It is a valuable substrate for the generation of both liquid and gaseous biofuels [46]. Some researchers examined the net energy input in bioethanol production and methane emissions from water hyacinth and switchgrass, using fermentation and anaerobic digestion [47]. Due to the higher output of water hyacinth, i.e., 60–100 tons/ha/yr compared to switchgrass which is 12.9 tons/ha/yr when chemical fertilizers are not used. Because of its accessible availability and high lingocellulosic content, water hyacinth meets the requirements as a formidable raw material for biofuel production [48]. A study where *Pichia stipitis NCIM 3497* was used to manufacture bioethanol (19.2 g/L) by utilizing water hyacinth [49]. In another investigation, the fermentation of this biomass by using *P. stipitis,* and *S. cerevisiae* yielded ethanol concentrations of 10.44, and 6.76 g/L, respectively [50]. Another study estimated that 9.62 metric tons of water hyacinth dry biomass can create 20.2 kg of sugar per day, which can make 1131.3 L of ethanol per day [51].

Duckweed is also a promising biomass for energy generation. With 37 species, it is one of the most abundant and tiniest plants on the planet [52]. Due to its resistance to high nutrient levels, it has also been frequently utilized in the processing of industrial and municipal wastewater [45]. Its specific growth rates are higher than those of other bigger aquatic plants. Landoltia Its biomass yielded 30.80.8 g/L of ethanol [53]. Duckweed was also discovered to be an excellent source of starch for bioethanol synthesis [54]. Researchers investigated the impact of high duckweed biomass loading concentrations on production of bioethanol, finding that high biomass loading (20% w/v) lowered the ethanol output to 18.8%, compared to an ethanol yield of 80% at low biomass loading [55]. The genus *Azolla* contains seven species that can be found in marshes, ponds, and ditches. It is rapid growing weed, with a biomass that doubles every 5–7 days [56]. Because of their lignocellulosic content, Azolla spp. is regarded a promising biofuel substrate. Azolla yielded 0.09 g/g of ethanol [57]. The fermentation of salvinia with *S. cerevisiae* and *S. carlsbergensis* yielded a 2.0 g/L ethanol yield [58].

Typha contains a variety of sugars in its leaf, stem, and root, making it an excellent source of ethanol. After doing both pretreatment and enzymatic digestion of Typha, the study investigated the highest glucose production (97% of the cellulose), with a theoretical ethanol yield of roughly 90%.

Water lettuce, or *Pistia stratiotes*, is a aquatic and poisonous plant with a growth degree similar to water hyacinth. Water lettuce has 49.45% carbohydrate, 16.47%

protein, 3.56% fat, and 17.81% crude fiber, indicating that it can be used to produce bioenergy. Using water lettuce as a feedstock, reported 14.9 g/L of ethanol.

4.6 Role of Nanotechnology for Advanced Ethanol Production

The biofuel sectors are currently concerned about high production costs and other technological hurdles. Nanotechnology is gaining popularity in this aspect, owing to environmental and economic concerns. Because it is useful and applied at both molecular and cellular levels, nanotechnology is the most important topic in modern science. Nanobiotechnology's uses in the bioenergy sector have grown in recent years. Nanoparticles have characteristics that distinguish them from bulk materials in that they are small enough to confine electrons and produce quantum effects [59].

The use of nanoparticles improves the performance of pretreatment, enzymatic saccharification, and fermentation processes. Particle size, shape, surface area, nature of nanoparticles, and kind of biomass used are all significant parameters for producing end-products and managing reaction rate control [60]. Because of their small size, metal nanoparticles can penetrate biomass cell walls.

Several studies have shown that metallic nanoparticles can be used as a cofactor to promote enzymatic stability and immobilization of enzymes onto a support material, resulting in increased enzymatic activity. Because of their ease of recovery and reusability, enzyme immobilization on nanomaterials lowers processing costs [59]. The efficiency of the process is improved by employing various nanoparticles, which give a wide immobilization surface for enzymes, as well as a longer self-life and stability [60]. According to a study, iron nanoparticles (Fe_3O_4) and nanocomposites (Fe_3O_4/alginate) effectively increased enzymatic activity and stability by providing optimal support for enzyme immobilization [61]. Co-precipitation was used to make Fe_3O_4 nanoparticles and specially structured nanocomposites for application in bioethanol production.

The immobilization of cellulase onto magnetic nanoparticles for bioethanol production was demonstrated [62]. Researchers covalently immobilized *Rhizopus oryzae* related lipase onto graphene oxide nanoparticles and showed that it had increased activity, solvent resistance, and temperature stability [63]. In reality, the covalent attachment of a specific enzyme to the support matrix increases the enzyme's self-life and reduces the processing cost. Large surface area for enzyme immobilization and catalytic site for ethanol oxidation is provided by graphene-based nanomaterials [64]. Enzyme immobilization on the graphene oxide surface can be achieved without the need for any cross-linking reagents or surface modifications and has no effect on the enzymes' heat or solvent resistance [63].

Iron oxides, cobalt oxides, copper oxides, manganese oxides, and other metallic nanoparticles operate as promising catalytic sources for the creation of renewable energy. In syngas fermentation, a study found that methyl-functionalized silica nanoparticles increased bioethanol output by 166.1% [65]. The scientists used a variety of nanoparticles in this study like palladium on carbon, silica, hydroxyl-functionalized single-walled carbon nanotubes, palladium on alumina, and iron oxide). The efficiency of silica nanoparticles for enhanced gas–liquid mass transfer was demonstrated, and the activity was further enhanced by adding hydrophobic functional groups (methyl and isopropyl). Researchers used methyl-functionalized magnetic nanoparticles to boost production of bioethanol during syngas fermentation, with methyl-functionalized cobalt-ferrite-silica ($CoFe_2O_4@SiO_2$-CH_3) nanoparticles producing roughly 213.5$ more [66].

A study has used alginate/magnetic nanoparticles that covalently bonded on chitosan-magnetite nanoparticles and cellulose-coated magnetic nanoparticles to exhibit improved bioethanol production [67]. The yeast cells were entrapped in a matrix of alginate/magnetic nanoparticles and mounted on magnetite-containing chitosan, yielding roughly 91% ethanol. For determining sugars production, a glass carbon electrode with graphene oxide constituting Cu nanoparticles was used and obtained greater accuracy and reusability [68]. Some created ultrathin 2D polycrystalline ZnO nanosheets with evenly scattered Ag nanoparticles to improve ethanol production surface reactions [69].

Magnetic nanoparticles were also used to trap cellulase, which improved the enzymatic saccharification of pretreated hemp biomass [70]. By fixing nanoparticles of iron oxide on the exterior of chitosan nanoparticles, created magnetic chitosan microspheres [71]. Cellulase from Aspergillus niger was immobilized on -cyclodextrin-coated magnetic nanoparticles [47]. In comparison to free cellulose, immobilized cellulose released a larger amount of glucose. The hydrolysis of jatropha, bagasse and plukenetia hulls with sulfonated magnetic carbonaceous acid nanoparticles demonstrated a considerable rate of bioconversion [72]. Cellulase immobilized on magnetic nanoparticles was used as a nano-bio catalyst to digest Sesbania aculeate biomass, yielding 5.31 g/L of bioethanol in another study [73]. Cellulase was immobilized on iron oxide nanoparticles made from Alternaria alternative cell filtrate, which demonstrated a high rate of cellulose conversion [74]. A group of researchers immobilized cellulase onto manganese oxide nanoparticles, increasing cellulase activity and providing greater support [75]. Over a wide temperature and pH range, cellulase immobilized on MnO2 nanoparticles degraded cellulosic compounds. Cellulase immobilized on MnO2 nanoparticles had strong cellulolytic activity, according to the findings. Cellulase was also used to make Ag and Au nanoparticles. The cellulase-assisted nanoparticles were also used as an immobilization matrix for cellulase [76].

When compared to free enzyme, the immobilized cellulase on produced Ag nanoparticles retained roughly 80% of its activity, according to thermal stability study. Scientists evaluated the influence of various metallic oxide nanoparticles on S. cerevisiae ethanol production in a separate investigation [77]. Higher quantities of nanoparticles employed resulted in lower ethanol concentrations. Fe3O4 nanoparticles boosted ethanol synthesis the most, with a maximum output of 0.26 g/g. Furthermore, adding NiO and Fe_3O_4 nanoparticles to the SSF method increased the yield of ethanol from potato peels by 1.6 and 1.13 times, respectively. Findings showed the optimal applications of NiO and Fe_3O_4 nanoparticles in bioethanol synthesis from agriculture wastes. Some optimized ethanol yields utilizing nickel oxide nanoparticles as a biocatalyst in a recent study [78]. The optimized procedure produced biomass concentrations of 2.04 g/L and ethanol yields of 0.26 g/g, respectively.

4.7 Global Scenario for Production of Ethanol in Coming Future

The global ethanol production is expected to rise from over 120 billion (bln) L in 2017 to nearly 131 billion (bln) L by 2027. In order to meet domestic demand, around half of the increase is likely to come from Brazil. China, India, Thailand, and the Philippines are the other major contributors to ethanol. US is likely to remain the leading ethanol producer, with Brazil, China, and the European Union following closely behind. The primary feedstocks for ethanol production are expected to remain coarse grains and sugarcane. In 2027, studies predict that 15 and 18% of worldwide maize and sugarcane production would be used for ethanol production, respectively. In the early years of the forecast period, ethanol production sourced primarily from maize in the United States should remain around 61.6 billion liters. With fewer local and international needs due to declining gasoline demand in developed nations, production of ethanol in the United States is expected to drop to 60.4 billion liters in the coming years. Brazil's ethanol production is expected to rise to 32.7 billion liters by 2027.

China's position as the world's third-largest ethanol producer should be solidified by 2027, with output hitting 11 billion liters. In China, ethanol is planned to be generated locally from maize and cassava using domestic stocks. By 2027, ethanol output from wheat, coarse grains, and sugar beet in the European Union is expected to drop to 7.1 billion liters. Sugar beet ethanol production is likely to stabilize at around 1.4 billion liters. In fact, due to greater production costs, sugar beet ethanol should be less lucrative in the European Union than ethanol made from other grain feedstocks. Thailand's ethanol production is expected to increase by roughly 6% every year. Although productivity has been primarily based on sugarcane, molasses, and cassava, sugarcane could increase its share of the market due to the limited availability of the other two feedstuffs to fulfill rising demand. Thailand's ethanol production is expected to reach 3.2 billion liters by 2027. India is forecast to raise ethanol production by 0.8 billion liters, with molasses accounting for nearly all of the overall output. Global ethanol consumption is expected to rise by around 12 billion liters, with 80% of this growth occurring in developing countries. Brazil's ethanol consumption is expected to increase by 5.4 billion liters, accounting for 42% of the global increase. Thailand has increased its ethanol use by 1 billion liters in the last few years. Ethanol demand in India is expected to grow at a 4.5% annual rate, adding 0.7 billion liters by 2027 compared to the baseline period. In the US, ethanol use is tied to a mandate in effect and constrained by a modestly growing blend wall as well as the potential of falling gasoline use.

References

1. Owusu, P. A., & Asumadu-Sarkodie, S. (2016). A review of renewable energy sources, sustainability issues and climate change mitigation. *Cogent Engineering, 3*(1), 1167990.
2. Park, J. Y., Shiroma, R., Al-Haq, M. I., Zhang, Y., Ike, M., Arai-Sanoh, Y., Ida, A., Kondo, M., & Tokuyasu, K. (2010). A novel lime pretreatment for subsequent bioethanol production from rice straw–calcium capturing by carbonation (CaCCO) process. *Bioresource Technology, 101*(17), 6805–6811.
3. Jeevan Kumar, S. P., Sampath Kumar, N. S., & Chintagunta, A. D. (2020). Bioethanol production from cereal crops and lignocelluloses rich agro-residues: Prospects and challenges. *SN Applied Sciences, 2*(10), 1–11.
4. Kalair, A., Abas, N., Saleem, M. S., Kalair, A. R., & Khan, N. (2021). Role of energy storage systems in energy transition from fossil fuels to renewables. *Energy Storage, 3*(1), e135.
5. Chandel, A. K., Garlapati, V. K., Jeevan Kumar, S. P., Hans, M., Singh, A. K., & Kumar, S. (2020). The role of renewable chemicals and biofuels in building a bioeconomy. *Biofuels, Bioproducts and Biorefining, 14*(4), 830–844. Sage, R. F. (1999). Why C4 photosynthesis. *C4 plant biology*, 3–16.
6. Banerjee, R., Chintagunta, A. D., & Ray, S. (2019). Laccase mediated delignification of pineapple leaf waste: An ecofriendly sustainable attempt towards valorization. *BMC Chemistry, 13*(1), 1–11.
7. Bušić, A., Marđetko, N., Kundas, S., Morzak, G., Belskaya, H., Ivančić Šantek, M., Komes, D., Novak, S., & Šantek, B. (2018). Bioethanol production from renewable raw materials and its separation and purification: A review. *Food Technology and Biotechnology, 56*(3), 289–311.
8. Aarti, C., Khusro, A., & Agastian, P. (2018). Carboxymethyl cellulase production optimization from Glutamicibacter arilaitensis strain ALA4 and its application in lignocellulosic waste biomass saccharification. *Preparative Biochemistry and Biotechnology, 48*(9), 853–866.
9. Anwar, Z., Gulfraz, M., & Irshad, M. (2014). Agro-industrial lignocellulosic biomass a key to unlock the future bio-energy: A brief review. *Journal of Radiation Research and Applied Sciences, 7*(2), 163–173.
10. Ballerini, D., Desmarquest, J. P., Pourquie, J., Nativel, F., & Rebeller, M. (1994). Ethanol production from lignocellulosics: Large scale experimentation and economics. *Bioresource Technology, 50*(1), 17–23.
11. Saini, J. K., Saini, R., & Tewari, L. (2015). Lignocellulosic agriculture wastes as biomass feedstocks for second-generation bioethanol production: Concepts and recent developments. *3 Biotech, 5*(4), 337–353.
12. Prasad, S., Singh, A., & Joshi, H. C. (2007). Ethanol as an alternative fuel from agricultural, industrial and urban residues. *Resources, Conservation and Recycling, 50*(1), 1–39.
13. Munasinghe, P. C., & Khanal, S. K. (2010). Biomass-derived syngas fermentation into biofuels: Opportunities and challenges. *Bioresource Technology, 101*(13), 5013–5022.
14. Demirbas, A. (2007). Progress and recent trends in biofuels. *Progress in Energy and Combustion Science, 33*(1), 1–18.
15. Wang, H., Wang, J., Fang, Z., Wang, X., & Bu, H. (2010). Enhanced bio-hydrogen production by anaerobic fermentation of apple pomace with enzyme hydrolysis. *International Journal of Hydrogen Energy, 35*(15), 8303–8309.
16. Marzialetti, T., Valenzuela Olarte, M. B., Sievers, C., Hoskins, T. J., Agrawal, P. K., & Jones, C. W. (2008). Dilute acid hydrolysis of Loblolly pine: A comprehensive approach. *Industrial & Engineering Chemistry Research, 47*(19), 7131–7140.

17. Dubois, J. L. (2011). Requirements for the development of a bioeconomy for chemicals. *Current Opinion in Environmental Sustainability, 3*(1–2), 11–14.
18. Brussels, Belgium 2019. Biomass for Energy—Agricultural Residues and Energy Crops; *BioEnergy_Europe: Factsheet*
19. IEA. (2010) *Sustainable Production of Second-Generation Biofuels*; IEA: Paris, France.
20. Smullen, E., Finnan, J., Dowling, D., & Mulcahy, P. (2017). Bioconversion of switchgrass: Identification of a leading pretreatment option based on yield, cost and environmental impact. *Renewable Energy, 111*, 638–645.
21. Hoadley, R. B. (2000). *Understanding wood: a craftsman's guide to wood technology*. Taunton press.
22. Neiva, D. M., Araujo, S., Gominho, J., de Cássia Carneiro, A., & Pereira, H. (2018). Potential of Eucalyptus globulus industrial bark as a biorefinery feedstock: Chemical and fuel characterization. *Industrial Crops and Products, 123*, 262–270.
23. Romaní, A., Larramendi, A., Yáñez, R., Cancela, Á., Sánchez, Á., Teixeira, J. A., & Domingues, L. (2019). Valorization of Eucalyptus nitens bark by organosolv pretreatment for the production of advanced biofuels. *Industrial Crops and Products, 132*, 327–335.
24. Zhao, J., Tian, D., Shen, F., Hu, J., Zeng, Y., & Huang, C. (2019). Valorizing waste lignocellulose-based furniture boards by phosphoric acid and hydrogen peroxide (Php) pretreatment for bioethanol production and high-value lignin recovery. *Sustainability, 11*(21), 6175.
25. Lamers, P., Searcy, E., Hess, J. R., & Stichnothe, H. (Eds.). (2016). *Developing the global bioeconomy: technical, market, and environmental lessons from bioenergy*. Academic.
26. Sarkar, N., Ghosh, S. K., Bannerjee, S., & Aikat, K. (2012). Bioethanol production from agricultural wastes: An overview. *Renewable Energy, 37*(1), 19–27.
27. Rojas-Chamorro, J. A., Romero, I., López-Linares, J. C., & Castro, E. (2020). Brewer's spent grain as a source of renewable fuel through optimized dilute acid pretreatment. *Renewable Energy, 148*, 81–90.
28. FAO. *Yearbook of Forest Products 2015*; Food and Agriculture Organization of the United Nations: London, UK, 2017; Vol. 2, pp. 397–467.
29. Pereira, S. R., & Sanchez i Nogue, V., Frazão, C.J., Serafim, L.S., Gorwa-Grauslund, M.F. and Xavier, A.M. (2015). Adaptation of Scheffersomyces stipitis to hardwood spent sulfite liquor by evolutionary engineering. *Biotechnology for Biofuels, 8*(1), 1–8.
30. Gottumukkala, L. D., Haigh, K., Collard, F. X., Van Rensburg, E., & Görgens, J. (2016). Opportunities and prospects of biorefinery-based valorisation of pulp and paper sludge. *Bioresource Technology, 215*, 37–49.
31. Schroeder, B. G., Zanoni, P. R. S., Magalhães, W. L. E., Hansel, F. A., & Tavares, L. B. B. (2017). Evaluation of biotechnological processes to obtain ethanol from recycled paper sludge. *Journal of Material Cycles and Waste Management, 19*(1), 463–472.
32. Nikolić, S., Lazić, V., Veljović, Đ, & Mojović, L. (2017). Production of bioethanol from pretreated cotton fabrics and waste cotton materials. *Carbohydrate Polymers, 164*, 136–144.
33. Keshav, P. K., Shaik, N., Koti, S., & Linga, V. R. (2016). Bioconversion of alkali delignified cotton stalk using two-stage dilute acid hydrolysis and fermentation of detoxified hydrolysate into ethanol. *Industrial Crops and Products, 91*, 323–331.
34. Wang, M., Zhou, D., Wang, Y., Wei, S., Yang, W., Kuang, M., Ma, L., Fang, D., Xu, S., & Du, S. K. (2016). Bioethanol production from cotton stalk: A comparative study of various pretreatments. *Fuel, 184*, 527–532.
35. Singh, J., Sharma, A., Sharma, P., Singh, S., Das, D., Chawla, G., Singha, A., & Nain, L. (2020). Valorization of jute (Corchorus sp.) biomass for bioethanol production. *Biomass Conversion and Biorefinery*, pp.1–12.
36. Eurostat. Municipal Waste Statistics. http://appsso.eurostat.ec.europa.eu/nui/submitViewTableAction.do
37. Tyagi, V. K., Fdez-Güelfo, L. A., Zhou, Y., Álvarez-Gallego, C. J., Garcia, L. R., & Ng, W. J. (2018). Anaerobic co-digestion of organic fraction of municipal solid waste (OFMSW): Progress and challenges. *Renewable and Sustainable Energy Reviews, 93*, 380–399.

38. Moreno, A. D., Magdalena, J. A., Oliva, J. M., Greses, S., Lozano, C. C., Latorre-Sánchez, M., Negro, M. J., Susmozas, A., Iglesias, R., Llamas, M., & Tomás-Pejó, E. (2021). Sequential bioethanol and methane production from municipal solid waste: An integrated biorefinery strategy towards cost-effectiveness. *Process Safety and Environmental Protection, 146*, 424–431.
39. Chanoca, A., De Vries, L., & Boerjan, W. (2019). Lignin engineering in forest trees. *Frontiers in Plant Science, 10*, 912.
40. Boerjan, W., Ralph, J., & Baucher, M. (2003). Lignin biosynthesis. *Annual Review of Plant Biology, 54*(1), 519–546.
41. Umezawa, T. (2018). Lignin modification in planta for valorization. *Phytochemistry Reviews, 17*(6), 1305–1327.
42. Park, J. J., Yoo, C. G., Flanagan, A., Pu, Y., Debnath, S., Ge, Y., Ragauskas, A. J., & Wang, Z. Y. (2017). Defined tetra-allelic gene disruption of the 4-coumarate: Coenzyme A ligase 1 (Pv4CL1) gene by CRISPR/Cas9 in switchgrass results in lignin reduction and improved sugar release. *Biotechnology for Biofuels, 10*(1), 1–11.
43. Yang, Y., Yoo, C. G., Guo, H. B., Rottmann, W., Winkeler, K. A., Collins, C. M., Gunter, L. E., Jawdy, S. S., Yang, X., Guo, H., & Pu, Y. (2017). Overexpression of a Domain of Unknown Function 266-containing protein results in high cellulose content, reduced recalcitrance, and enhanced plant growth in the bioenergy crop Populus. *Biotechnology for Biofuels, 10*(1), 1–13.
44. Vučurović, V. M., & Razmovski, R. N. (2012). Sugar beet pulp as support for Saccharomyces cerivisiae immobilization in bioethanol production. *Industrial Crops and Products, 39*, 128–134.
45. Sindhu, R., Binod, P., Pandey, A., Madhavan, A., Alphonsa, J. A., Vivek, N., Gnansounou, E., Castro, E., & Faraco, V. (2017). Water hyacinth a potential source for value addition: An overview. *Bioresource technology, 230*, 152–162.
46. Huang, W. (2015). An integrated biomass production and conversion process for sustainable bioenergy. *Sustainability, 7*(1), 522–536.
47. Rezania, S., Ponraj, M., Din, M. F. M., Songip, A. R., Sairan, F. M., & Chelliapan, S. (2015). The diverse applications of water hyacinth with main focus on sustainable energy and production for new era: An overview. *Renewable and Sustainable Energy Reviews, 41*, 943–954.
48. Magdum, S., More, S., & Nadaf, A. (2012). Biochemical conversion of acid-pretreated water hyacinth (Eichhornia Crassipes) to alcohol using Pichia Stipitis NCIM3497. *International Journal of advanced biotechnology and research, 3*(2), 585–590.
49. Das, S., Bhattacharya, A., Haldar, S., Ganguly, A., Gu, S., Ting, Y. P., & Chatterjee, P. K. (2015). Optimization of enzymatic saccharification of water hyacinth biomass for bio-ethanol: Comparison between artificial neural network and response surface methodology. *Sustainable Materials and Technologies, 3*, 17–28.
50. Malveaux, C. C. (2013). Coastal plants for biofuel production and coastal preservation.
51. Xu, J., Cui, W., Cheng, J. J., & Stomp, A. M. (2011). Production of high-starch duckweed and its conversion to bioethanol. *Biosystems Engineering, 110*(2), 67–72.
52. Perniel, M., Ruan, R., & Martinez, B. (1998). Nutrient removal from a stormwater detention pond using duckweed. *Applied Engineering in Agriculture, 14*(6), 605–609.
53. Cheng, J. J., & Stomp, A. M. (2009). Growing duckweed to recover nutrients from wastewaters and for production of fuel ethanol and animal feed. *Clean-Soil, Air, Water, 37*(1), 17–26.
54. Zhao, X., Moates, G. K., Elliston, A., Wilson, D. R., Coleman, M. J., & Waldron, K. W. (2015). Simultaneous saccharification and fermentation of steam exploded duckweed: Improvement of the ethanol yield by increasing yeast titre. *Bioresource Technology, 194*, 263–269.
55. Kollah, B., Patra, A. K., & Mohanty, S. R. (2016). Aquatic microphylla Azolla: A perspective paradigm for sustainable agriculture, environment and global climate change. *Environmental Science and Pollution Research, 23*(5), 4358–4369.
56. Miranda, A. F., Biswas, B., Ramkumar, N., Singh, R., Kumar, J., James, A., Roddick, F., Lal, B., Subudhi, S., Bhaskar, T., & Mouradov, A. (2016). Aquatic plant Azolla as the universal feedstock for biofuel production. *Biotechnology for biofuels, 9*(1), 1–17.

57. Abdullahi, A. F., Maikaje, D. B., Denwe, S. D., & Muhammad, M. N. (2016). Evaluation of fermentation products of Eichhornia crassipes, Pistia stratiotes and Salvinia molesta. *Agriculture and Biology Journal of North America, 7*(1), 27–31.
58. Mohamad, N. R., Marzuki, N. H. C., Buang, N. A., Huyop, F., & Wahab, R. A. (2015). An overview of technologies for immobilization of enzymes and surface analysis techniques for immobilized enzymes. *Biotechnology & Biotechnological Equipment, 29*(2), 205–220.
59. Kim, J., Grate, J. W., & Wang, P. (2006). Nanostructures for enzyme stabilization. *Chemical engineering science, 61*(3), 1017–1026.
60. Srivastava, N., Singh, J., Ramteke, P. W., Mishra, P. K., & Srivastava, M. (2015). Improved production of reducing sugars from rice straw using crude cellulase activated with Fe_3O_4/Alginate nanocomposite. *Bioresource technology, 183*, 262–266.
61. Jordan, J., Kumar, C. S., & Theegala, C. (2011). Preparation and characterization of cellulase-bound magnetite nanoparticles. *Journal of Molecular Catalysis B: Enzymatic, 68*(2), 139–146.
62. Hermanová, S., Zarevúcká, M., Bouša, D., Pumera, M., & Sofer, Z. (2015). Graphene oxide immobilized enzymes show high thermal and solvent stability. *Nanoscale, 7*(13), 5852–5858.
63. Kakaei, K., Rahimi, A., Husseindoost, S., Hamidi, M., Javan, H., & Balavandi, A. (2016). Fabrication of Pt–CeO_2 nanoparticles supported sulfonated reduced graphene oxide as an efficient electrocatalyst for ethanol oxidation. *International Journal of Hydrogen Energy, 41*(6), 3861–3869.
64. Kim, Y. K., Park, S. E., Lee, H., & Yun, J. Y. (2014). Enhancement of bioethanol production in syngas fermentation with Clostridium ljungdahlii using nanoparticles. *Bioresource Technology, 159*, 446–450.
65. Kim, Y. K., & Lee, H. (2016). Use of magnetic nanoparticles to enhance bioethanol production in syngas fermentation. *Bioresource Technology, 204*, 139–144.
66. Ivanova, V., Petrova, P., & Hristov, J. (2011). Application in the ethanol fermentation of immobilized yeast cells in matrix of alginate/magnetic nanoparticles, on chitosan-magnetite microparticles and cellulose-coated magnetic nanoparticles. arXiv preprint arXiv:1105.0619.
67. Santos, F. C. U., Paim, L. L., da Silva, J. L., & Stradiotto, N. R. (2016). Electrochemical determination of total reducing sugars from bioethanol production using glassy carbon electrode modified with graphene oxide containing copper nanoparticles. *Fuel, 163*, 112–121.
68. Lin, L., Liu, T., Zhang, Y., Liang, X., Sun, R., Zeng, W., & Wang, Z. (2016). Enhancing ethanol detection by heterostructural silver nanoparticles decorated polycrystalline zinc oxide nanosheets. *Ceramics International, 42*(2), 3138–3144.
69. Abraham, R. E., Verma, M. L., Barrow, C. J., & Puri, M. (2014). Suitability of magnetic nanoparticle immobilised cellulases in enhancing enzymatic saccharification of pretreated hemp biomass. *Biotechnology for biofuels, 7*(1), 1–12.
70. Miao, X., Pi, L., Fang, L., Wu, R., & Xiong, C. (2016). Application and characterization of magnetic chitosan microspheres for enhanced immobilization of cellulase. *Biocatalysis and Biotransformation, 34*(6), 272–282.
71. Su, T. C., Fang, Z., Zhang, F., Luo, J., & Li, X. K. (2015). Hydrolysis of selected tropical plant wastes catalyzed by a magnetic carbonaceous acid with microwave. *Scientific Reports, 5*(1), 1–14.
72. Baskar, G., Kumar, R. N., Melvin, X. H., Aiswarya, R., & Soumya, S. (2016). Sesbania aculeate biomass hydrolysis using magnetic nanobiocomposite of cellulase for bioethanol production. *Renewable Energy, 98*, 23–28.
73. Ingle, A. P., Rathod, J., Pandit, R., da Silva, S. S., & Rai, M. (2017). Comparative evaluation of free and immobilized cellulase for enzymatic hydrolysis of lignocellulosic biomass for sustainable bioethanol production. *Cellulose, 24*(12), 5529–5540.
74. Cherian, E., Dharmendirakumar, M., & Baskar, G. (2015). Immobilization of cellulase onto MnO_2 nanoparticles for bioethanol production by enhanced hydrolysis of agricultural waste. *Chinese Journal of Catalysis, 36*(8), 1223–1229.
75. Mishra, A., & Sardar, M. (2015). Cellulase assisted synthesis of nano-silver and gold: Application as immobilization matrix for biocatalysis. *International Journal of Biological Macromolecules, 77*, 105–113.

76. Sanusi, I. A., Faloye, F. D., & Gueguim Kana, E. B. (2019). Impact of various metallic oxide nanoparticles on ethanol production by Saccharomyces cerevisiae BY4743: Screening, kinetic study and validation on potato waste. *Catalysis Letters, 149*(7), 2015–2031.
77. Sanusi, I. A., Suinyuy, T. N., Lateef, A., & Kana, G. E. (2020). Effect of nickel oxide nanoparticles on bioethanol production: Process optimization, kinetic and metabolic studies. *Process Biochemistry, 92*, 386–400.

Chapter 5
Potential of Weed Biomass for Bioethanol Production

Abstract Bioethanol can be considered as future fuel which has the potential to replace fossil fuels. Weeds have become difficult to manage to stop their global spread. Weeds can be profitable and valuable when utilized as a potential resource. Next-generation biofuel feedstocks can be made from weeds because of their rapid reproduction, abundance of cellulose, and low lignin content. Pretreatment of lignocellulosic materials is one of the most important aspects of a cost-effective bioethanol production process. This chapter provides an overview of weed biomass, several pretreatment methods for LCB, and the fermentation process for producing bioethanol from lignocellulosic weed. The current chapter discusses bioethanol production from weed biomass.

Keywords Bioethanol · Weed · Lignocelluloses · Fermentation

5.1 Introduction

As the world's population has risen and industrialization has progressed, the rate of energy consumption has increased. The world is currently dealing with a massive issue of fossil-fuel depletion, while energy consumption is increasing at an alarming rate. The design and deployment of technology based on alternative energy sources is one feasible answer to this problem. Bioenergy is a long-term strategy for reducing reliance on fossil fuels. Lignocellulosic biomass (e.g., grasses, weeds, industrial and municipal waste) is being studied in depth for the production of 2G bioethanol. The majority of biofuels, like bioethanol, biogas, biodiesel, and biohydrogen, are produced from agricultural wastes, which helps to alleviate environmental pressure. Bioenergy resources like starch and cellulose are important because they are cost-effective and environmentally friendly. To avoid the problems of agricultural land and food security, several research studies have lately explored the use of weed biomass as an alternative feedstock for bioethanol production. They can very well survive in harsh environmental circumstances which are harmful to other plant growth. As a result, biofuels can be produced from them at any time of year. Many weedy plant species can be found all over the world. They produce a lot of dry matter with very little water and nutrients, making them highly resistant and able to survive in any

N. K. Aggarwal et al., *Bioethanol Production*, Green Chemistry and Sustainable Technology, https://doi.org/10.1007/978-3-031-05091-6_5

climatic and edaphic condition [1]. As a feedstock for biodiesel production, aquatic weed biomass has several benefits over terrestrial energy crops. Biomass from aquatic weeds does not always necessitate precious resources like cropland and freshwater.

The rapid growth of aquatic weeds means that their biomass production may be higher than that of many land-based energy crops. Aquatic weeds' global spread encourages their usage as a biofuel source to provide energy security. Weedy plants are a very desirable feedstock for bioethanol production because of their rich cellulose content and widespread distribution. The objective of this chapter is to provide an update on the current status of weeds' bioenergy potential, taking into account their vast accessibility, and high yield. This chapter significantly evaluates the potential of weeds as a feedstock for bioethanol production.

5.2 The Chemical Makeup of Lignocellulosic Weeds Biomass

The LCB is the most prevalent biopolymer on the planet, accounting for almost half of all biomass. Cellulose, hemicellulose, and lignin are the three main components of weed biomass. Cellulose and hemicelluloses together are known as holocellulose. (Fig. 5.1).

Cellulose is a crystalline, linear structure composed of glucose molecules linked together by β-1–4 glycosidic linkages. As a result of these linkages, cellulose has a very high crystalline structure and is resistant to breakdown. It is the world's most abundant organic polymer. Hemicellulose, on the other hand, is a linear and highly

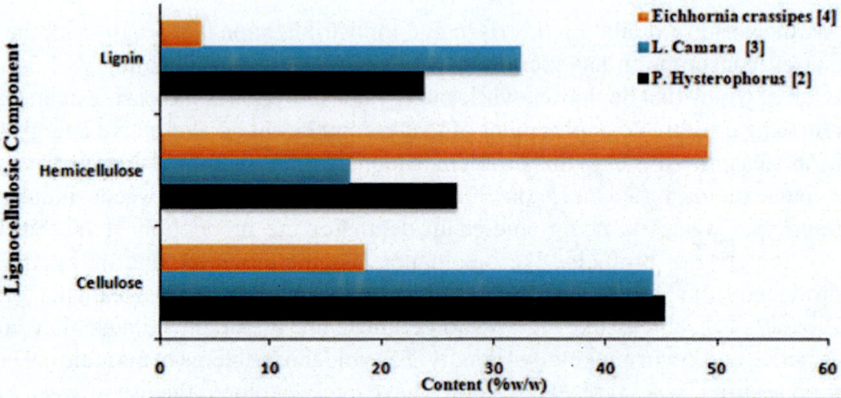

	Cellulose	Hemicellulose	Lignin
Eichhornia crassipes [4]	18.4	49.2	3.55
L. Camara [3]	44.1	17	32.25
P. Hysterophorus [2]	45.2	26.6	23.6

Fig. 5.1 Composition analysis of various lignocellulosic weeds biomass

branching combination of pentoses and hexoses. Lignin is a polyphenolic polymer with a high branching structure that gives biomass structural stability [2]. The major sugars in LCB are C_5 and C_6. The sugar component can be fermented directly to yield bioethanol. The three primary processes in the production of bioethanol from lignocellulosic materials are pretreatment/delignification of biomasses to generate cellulose, enzymatic breakdown of cellulose to sugars, and fermentation of sugars to bioethanol. Biomass pretreatment—Pretreatment is an important step in changing the structural properties of biomass and improving the accessibility of cellulose and hemicellulose to enzymes.

The goal of biomass pretreatment is to separate cellulose and hemicellulose from biomass and then hydrolyze them into fermentable sugar that can be fermented into bioethanol. The appearance of recalcitrant lignin on the cell wall surface prevents cellulose and hemicelluloses from being completely converted to ethanol [3, 4]. The lignocellulosic material's major structural component is cellulose, which has a high degree of crystallinity. The pretreatment disturbs the lignocellulosic matrix, exposing the subsurface polymers and reducing their crystallinity, increasing polysaccharide surface area. The effectiveness of a pretreatment technique is determined by several factors such as:

- Low operational and capital costs
- Minimal biomass pre-processing requirements
- Efficacy across a wide variety of biomass types and at high biomass loading
- Production of small numbers of fermentation inhibitors
- Chemical/catalyst recycling
- Limited waste generation
- Minimal corrosion
- Maximal breakdown with selective component removal
- Production of higher-value lignin co-products
- High hemicellulose recovery [5, 6].

Because it affects the efficiency and economy of downstream operations, pretreatment is the most important barrier in commercialized bioethanol production. Only around 20% of biomass is hydrolyzed in saccharification processes without pretreatment [7]. Alkali pretreatments increase cellulose digestibility and lignin solubilization while having a lower impact on cellulose and hemicellulose than acid and hydrothermal treatments [8]. The amount of lignin in the biomass determines how effective this approach is. The effect of an oxidizing agent like H_2O_2 alone or in combination with alkali or ammonia on biomass pretreatment is substantial. By promoting lignin removal, adding an oxidant agent (oxygen/H_2O_2) to an alkaline pretreatment ($NaOH/Ca(OH)_2$) can enhance performance [8]. Biological pretreatments use brown, white, and soft rot fungi, which degrade lignin and hemicellulose but not cellulose, which is more resistant than the other components [9]. Brown fungus pretreatment has recently been identified as a promising approach for increasing the enzymatic hydrolysis yields of *Pinus radiata* and *Pinus sylvestris*, with saccharification yields of approximately 70% [10]. Acid pretreatment, one of the most promising approaches, has been thoroughly investigated. It mostly results in hemicellulose solubilization, while it is less effective at removing lignin [11]. Concentrated or dilute acid can be used for acid pretreatment. Pretreatment has an effect on various biomasses depending on the process used and the type of LCB used. To make bioethanol from weed biomass, developing effective pretreatment conditions is essential.

5.3 Enzymatic Hydrolysis

After pretreatment of LCB, acid or enzymatic hydrolysis is employed to break down cellulose and sometimes hemicellulose into monomeric sugars like C_6 and C_5 [2]. Therefore, an essential stage in the production of bioethanol is the optimization of enzyme hydrolysis. Enzyme loadings, hydrolysis time, and physical and chemical properties of the pretreated substrate have a major effect on the rates and amount of enzyme hydrolysis of lignocellulosic materials. Degree of polymerization, crystallinity, accessible area and the availability of hemicellulose and lignin are all factors that influence enzymatic hydrolysis. Because it requires only mild reaction conditions (pH 4.8–5.0 and a temperature range of 45–50 °C), it does not promote reactor corrosion, and it produces few byproducts while yielding significant sugar yields, enzymatic hydrolysis is an excellent method for the conversion of cellulose into glucose. *Clostridium, Cellulomonas, Thermonospora, Bacillus, Bacteriodes, Ruminococcus, Erwinia, Acetovibrio, Microbispora, Streptomyces*, and other fungi such as *Trichoderma, Penicillium, Fusarium, Phanerochaete, Humicola, Schizophillum* sp. are among the cellulolytic microorganisms that produce cellulose enzyme. These enzymes can break down cellulose into glucose and galactose monomers. In comparison to acid or alkaline hydrolysis, enzymatic hydrolysis is advantageous due to its low toxicity, cheap utility cost, and low corrosion [12]. When using enzymatic hydrolysis, various levels of process integration are possible, such as separate, consolidated bioprocessing (CBP), separate hydrolysis and fermentation (SHF), simultaneous saccharification and fermentation (SSF), simultaneous saccharification and co-fermentation (SSCF) [13] (Fig. 5.2).

5.4 Fermentation

The hydrolysate produced by enzymatic saccharification of LCB has mostly been used as a fermentation substrate in the bioethanol production process. After enzymatic hydrolysis, the supernatant containing different sugars (C5 and C6) is fermented to bioethanol. Fermenting sugars from weed biomass to ethanol can be done with a number of microorganisms, including fungus and bacteria. The type of fermentation process chosen is determined by the properties of the biomass, enzymes, and microbes used. *Saccharomyces cerevisiae* and *Pichia stipitis* are natural microorganisms that can convert glucose and xylose to ethanol. As a result, conversion of

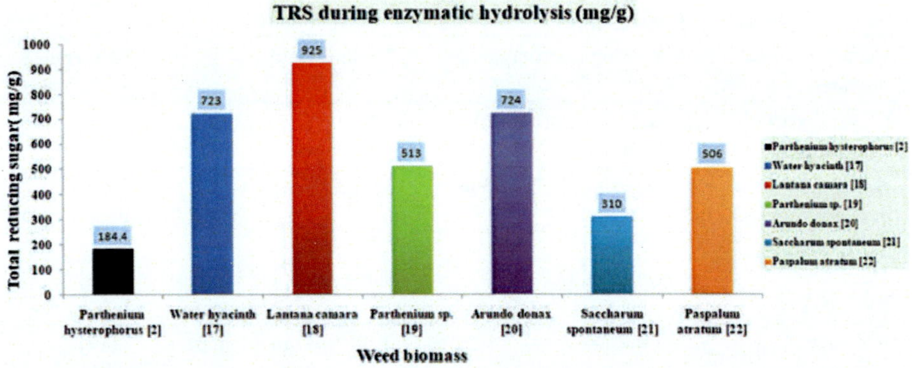

Fig. 5.2 Total reducing sugar released from lignocellulosic weed biomass after enzymatic hydrolysis

all types of sugars released from cellulose and hemicellulose is required for effective and economically viable ethanol synthesis from lignocellulose [14, 15]. Microbes, on the other hand, have a natural affinity for the carbohydrates they consume. The reaction of strains to inhibitors in fermenting media and various pretreatment procedures is used to select suitable strains for fermentation. For alkali and acid-pretreated substrates, *S. cerevisiae* JRC6 and *Candida tropicalis* JRC1 are indicated, respectively [16]. Kuhad et al. used SHF of *L. camara* to produce ethanol, with a highest bioethanol output of 148 g/1000 g of pretreated *L. camara* biomass [17]. After being pretreated with 4% H_2SO_4, water hyacinth biomass was saccharified with commercial cellulase enzyme (30FPU/g biomass) and fermented with *S. cerevisiae*, resulting in an ethanol concentration of 0.292% (w/v) and a process efficiency of 59.3% [18]. After fermenting hydrolysate from NaOH pretreated *Pennisetum polystachion* with *S. cerevisiae* (TISTR 5596), Prasertwasu [19] found a good bioethanol yield after 24 h. Another study utilizing a traditional strain, *S. cerevisiae* NBRC 2346, demonstrated a high yield of 14.9 g/1 ethanol from water lettuce [20] (Table 5.1).

5.5 Summary

Lignocellulosic or 2G bioethanol is being studied as a long-term, environmentally friendly alternative to depleting crude oil sources. The type of substrate chosen for the liberation of reducing sugars at competitive pricing influences the pretreatment method chosen. However, due to the variability of lignocellulose biomass, it is impossible to pick any one pretreatment as the optimum. The decision will be determined by the nature or source of the lignocellulosic to be treated, as well as the intended use of the hydrolysate. Bioethanol can be produced by fermenting sucrose or starch-containing plants. Bioethanol production on a huge scale, on the other

Table 5.1 Bioethanol production from lignocellulosic weed biomass

Lignocellulosic biomass	Fermenting microorganism	Ethanol yield	Reference
Carrot grass (*P. hysterophorus*)	*S. cerevisiae* MTCC170	0.15 g/g raw biomass	[21]
Panicum maximum cv. TD 53	*S. cerevisiae* TISTR 5596	5.9 g/L	[22]
Saccharum spontaneum	*P. stipitis* NCIM3498	0.40 ± 0.01 g/g biomass	[23]
Red sage (*L. camara*)	*S. cerevisiae*	0.148 g/g raw biomass	[17]
E. crassipes	*S. cerevisiae*	4.4 gL^{-1}	[24]
water hyacinth	*P. stipitis* NCIM 3497	19.2 gL^{-1}	[25]
Lemna minor	*S. cerevisiae*	0.218 g/g biomass	[26]

hand, necessitates the utilization of very vast agricultural areas for maize or sugarcane production. Bioethanol commercialization is hampered by a lack of feedstock and a lack of long-term viability. Due to its abundant availability, weed biomass for the production of bioethanol appears to be a solution to this problem. Therefore, weed biomass can be used as an alternative feedstock and is cost-effective for fuel bioethanol production.

References

1. Priya, H. R., Pavithra, A. H., & Divya, J. (2014). Prospects and problems of utilization of weed biomass: a review. *Research & Reviews: Journal of Agriculture and Allied Sciences, 3*(2).
2. Banka, A., Komolwanich, T., & Wongkasemjit, S. (2015). Potential Thai grasses for bioethanol production. *Cellulose, 22*(1), 9–29.
3. Kumar, P., Barrett, D. M., Delwiche, M. J., & Stroeve, P. (2009). Methods for pretreatment of lignocellulosic biomass for efficient hydrolysis and biofuel production. *Industrial & Engineering Chemistry Research, 48*(8), 3713–3729.
4. Mood, S. H., Golfeshan, A. H., Tabatabaei, M., Jouzani, G. S., Najafi, G. H., Gholami, M., & Ardjmand, M. (2013). Lignocellulosic biomass to bioethanol, a comprehensive review with a focus on pretreatment. *Renewable and Sustainable Energy Reviews, 27*, 77–93.
5. Sun, Y., & Cheng, J. (2002). Hydrolysis of lignocellulosic materials for ethanol production: A review. *Bioresource Technology, 83*(1), 1–11.
6. Taherzadeh, M. J., & Karimi, K. (2008). Pretreatment of lignocellulosic wastes to improve ethanol and biogas production: A review. *International Journal of Molecular Sciences, 9*(9), 1621–1651.
7. Bensah, E. C., & Mensah, M. (2013). Chemical pretreatment methods for the production of cellulosic ethanol: technologies and innovations. *International Journal of Chemical Engineering, 2013.*
8. Carvalheiro, F., Duarte, L. C., Gírio, F. M. (2008). Hemicellulose biorefineries: A review on biomass pretreatments.

9. Sánchez, C. (2009). Lignocellulosic residues: Biodegradation and bioconversion by fungi. *Biotechnology Advances, 27*(2), 185–194.
10. Ray, M. J., Leak, D. J., Spanu, P. D., & Murphy, R. J. (2010). Brown rot fungal early stage decay mechanism as a biological pretreatment for softwood biomass in biofuel production. *Biomass and Bioenergy, 34*(8), 1257–1262.
11. de Carvalho, D. M., Sevastyanova, O., Penna, L. S., da Silva, B. P., Lindström, M. E., & Colodette, J. L. (2015). Assessment of chemical transformations in eucalyptus, sugarcane bagasse and straw during hydrothermal, dilute acid, and alkaline pretreatments. *Industrial Crops and Products, 73*, 118–126.
12. Sarkar, N., Ghosh, S. K., Bannerjee, S., & Aikat, K. (2012). Bioethanol production from agricultural wastes: An overview. *Renewable Energy, 37*(1), 19–27.
13. Conde-Mejia, C., Jimenez-Gutierrez, A., & El-Halwagi, M. (2012). A comparison of pretreatment methods for bioethanol production from lignocellulosic materials. *Process Safety and Environmental Protection, 90*(3), 189–202.
14. Jeffries, T. W. (2006). Engineering yeasts for xylose metabolism. *Current Opinion in Biotechnology, 17*(3), 320–326.
15. Hahn-Hägerdal, B., Karhumaa, K., Jeppsson, M., & Gorwa-Grauslund, M. F., (2007). Metabolic engineering for pentose utilization in Saccharomyces cerevisiae. *Biofuels,* 147–177.
16. Wongwatanapaiboon, J., Kangvansaichol, K., Burapatana, V., Inochanon, R., Winayanuwattikun, P., Yongvanich, T., & Chulalaksananukul, W. (2012). The potential of cellulosic ethanol production from grasses in Thailand. *Journal of Biomedicine and Biotechnology, 2012.*
17. Kuhad, R. C., Gupta, R., Khasa, Y. P., & Singh, A. (2010). Bioethanol production from Lantana camara (red sage): Pretreatment, saccharification and fermentation. *Bioresource Technology, 101*(21), 8348–8354.
18. Satyanagalakshmi, K., Sindhu, R., Binod, P., Janu, K. U., Sukumaran, R. K., & Pandey, A. (2011). Bioethanol production from acid pretreated water hyacinth by separate hydrolysis and fermentation.
19. Prasertwasu, S., Khumsupan, D., Komolwanich, T., Chaisuwan, T., Luengnaruemitchai, A., & Wongkasemjit, S. (2014). Efficient process for ethanol production from Thai Mission grass (Pennisetum polystachion). *Bioresource Technology, 163*, 152–159.
20. Mishima, D., Kuniki, M., Sei, K., Soda, S., Ike, M., & Fujita, M. (2008). Ethanol production from candidate energy crops: water hyacinth (Eichhornia crassipes) and water lettuce (Pistia stratiotes L.). *Bioresource Technology, 99*(7), 2495–2500.
21. Singh, S., Agarwal, M., Sarma, S., Goyal, A., & Moholkar, V. S. (2015). Mechanistic insight into ultrasound induced enhancement of simultaneous saccharification and fermentation of Parthenium hysterophorus for ethanol production. *Ultrasonics Sonochemistry, 26*, 249–256.
22. Ratsamee, S., Akaracharanya, A., Leepipatpiboon, N., Srinorakutara, T., Kitpreechavanich, V., & Tolieng, V. (2012). Purple guinea grass: Pretreatment and ethanol fermentation. *BioResources, 7*(2), 1891–1906.
23. Chandel, A. K., Singh, O. V., Rao, L. V., Chandrasekhar, G., & Narasu, M. L. (2011). Bioconversion of novel substrate Saccharum spontaneum, a weedy material, into ethanol by Pichia stipitis NCIM3498. *Bioresource Technology, 102*(2), 1709–1714.
24. Aswathy, U. S., Sukumaran, R. K., Devi, G. L., Rajasree, K. P., Singhania, R. R., & Pandey, A. (2010). Bio-ethanol from water hyacinth biomass: An evaluation of enzymatic saccharification strategy. *Bioresource Technology, 101*(3), 925–930.
25. Magdum, S., More, S., & Nadaf, A. (2012). Biochemical conversion of acid-pretreated water hyacinth (Eichhornia Crassipes) to alcohol using Pichia Stipitis NCIM3497. *International Journal of advanced biotechnology and research, 3*(2), 585–590.
26. Gusain, R., & Suthar, S. (2017). Potential of aquatic weeds (Lemna gibba, Lemna minor, Pistia stratiotes and Eichhornia sp.) in biofuel production. *Process Safety and Environmental Protection, 109*, 233–241.

Chapter 6
Valorization of Organic Fraction of MSW for Bioethanol Production

Abstract Municipal solid waste management is a big problem all over the world. The improper management of municipal solid waste (MSW) can be extremely dangerous. Researchers from all around the world are working on a plethora of methods to reduce the generation of waste and oversee waste management for socioeconomic and environmental welfare. Using holistic and integrated methodologies, value-added goods can be made from municipal garbage. The framework for analyzing municipal solid waste to energy conversion and waste-derived bioeconomy in order to meet sustainable development goals is very important. In the context of the circular economy, it is critical to valorize the organic fraction of municipal solid waste (OFMSW). This chapter aims to look the state-of-the-art of valorization of OFMSW in bioethanol production. Further, different challenges and future aspects are also discussed.

Keywords Valorization · OFMSW · Biorefinery · Waste management · Bioethanol · Circular economy

6.1 Introduction

One of the major issues confronting humanity is the rise in solid waste output as a result of rising urban populations and living conditions [1]. Bioethanol generation from MSW has enormous potential in nations where energy security and waste management are increasingly the most pressing challenges [2]. This technology is still in the early stages of commercialization, and the expense of performing it on a large-scale is prohibitively expensive. Municipal wastes, hazardous wastes, clinical wastes, and radioactive wastes are the four types of solid waste. MSW is defined as any waste generated by different sectors like industrial, domestic, or possibly institutional establishments [3]. The amount of waste produced is directly proportional to a country's economic progress and socioeconomic level. It is also influenced by a variety of factors such as financial advancement, social standards, topographical area, and energy supplies [4]. There is a difference in the amount of biological waste generated by high-income, middle-income, and low-income countries' MSW, where biological waste encompasses the majority of MSW except metal, paper, plastic, and glass [5].

N. K. Aggarwal et al., *Bioethanol Production*, Green Chemistry and Sustainable Technology, https://doi.org/10.1007/978-3-031-05091-6_6

Food wastes, clothing, paper, glass, and plastic are the most common types of MSW, along with other hazardous family wastes including electric lights, batteries, discarded medications, and car components. MSW is viewed as originating primarily from family units, but it also includes trash from workplaces, lodgings, shopping edifices/stores, schools, organizations, and municipal management, such as road cleaning [6]. Other wastes, such as modern waste material, municipal wastewater, mechanical wastewater, stormwater, and hazardous waste, are often overlooked, revealing the intricacies and magnitude of the challenges [7].

MSWM is one of the world's most pressing issues in terms of collection, storage, transportation, and disposal. Garbage treatment can include everything from lowering the size of the waste to recovering energy from it. Different variables such as social, financial, legal, institutional, and environmental factors can be used to analyze the viability of sustainable waste management [8]. There is an urgent need to efficiently manage and eliminate MSW. The MSW life cycle assessment (LCA) takes into account all aspects of MSW, from trash collection to energy production (end-use). LCA is thought to be a viable method for avoiding compromises between economic and environmental factors. The economic and environmental impact of MSW such as valorization, prevention, and management should be evaluated using the analysis tool of LCA methods. MSW is biodegradable to the extent that it can be used as a feedstock for the production of bioenergy, biofuels, and product synthetic chemicals [9]. SWM has been a major concern around the world in the last 30 years. Sufficient structure for appropriate MSW management necessitates the participation of several growing nations as well as budgetary resources [10]. Different sources of MSW can be seen in Fig. 6.1.

Fig. 6.1 Different types of waste constituting MSW

6.2 Current Status of MSW

Concentrating on MSW has a greater global environmental impact. The typical person's contribution makes a significant difference. On an average, a person generates roughly 34% of global garbage during their lifetime, which is a huge amount when you consider the world's population of 3.37 billion people. Global trash creation is expected to reach 3.40 billion by 2050, according to estimates. India, the United States, China, Brazil, and Indonesia are the top MSW producers [11]. According to a detailed research published by the World Bank, countries with higher income typically produce 96% of waste, while countries having above-average income produce roughly 82%. Countries with a lower-than-average income and those with a very lower capital income generate trash at rates of 51% and 39%, respectively. The largest significant contributor to MSW is food waste [12, 13]. The European countries create the majority of food waste, accounting for over 69% of the 130 million tons of MSW produced annually [14]. Food waste accounts for 55% of China's annual MSW output of 170 million tons. Despite the fact that India's total municipal waste is 50 million tons per year, it contributes 40–60% of all food waste. According to the survey, food accounts for 35% of all municipal garbage [15]. The outcomes of a poorly managed waste management system could have a negative impact on the country's economy, human health, ecological degradation, biodiversity loss, and pollution of water, soil, and air. Separation (sorting) of sources, collection at preliminary level, site sweeping, trash storage, again collection at secondary level, transportation, process facility, disposal, participation of community, and population awareness are all sequential techniques in current waste management [16]. For the following reasons, waste management should be improved: The most pressing concern is environmental harm, with the goal of reducing the detrimental impact of flies and vermin manifestations on human health [17]. The factors of economic growth, resource utilization, and waste production are all interlinked. Landfills, recycling into new products, biological and thermal treatment are the four types of waste management now available. The majority of the global scenario deals with MSW management through landfills. According to statistics, only 15% of the garbage collected is recycled, and the rest is either disposed of in landfills or dumped in open areas, posing serious health risks. Carbon footprinting has been linked to the disposal of solid waste in landfills. The UN climate conference in 2014 confirmed that practice of landfilling MSW is major contributor to global methane emissions. In spite the fact that methane contributes to the atmosphere, ordinary carbon dioxide contributes roughly 800 tons per year to the atmosphere. Methane produced from MSW contributes approximately 5% to global warming, and that is 21–25 times larger than carbon dioxide. Modern management practices of municipal waste should be sanitary and with continual monitoring of disposal of food waste. Try to incorporate creative methods for making energy from waste, which can be useful in reducing waste. The technique of generating electricity from waste can improve a country's economic development while simultaneously lowering pollution. Bio-wastes are made up of the different components like: cellulose, hemicellulose, lipids, and proteins in general. These units serve as a feedstock for production of energy from waste, reducing pollution indirectly since when the solid wastes are not targeted under waste management and instead disposed of in landfills, damaging gases such as methane and carbon dioxide are released, contributing to global warming [18]. Biological and physiochemical processes can be used to generate energy from trash. A real-world worldwide example of energy conversion from garbage can be explored. Finland ranks top in all categories, including MSW management, according to The Good Country Index. Between 2003 and 2017, there has been a significant decrease in landfilling and a significant rise in both material recovery and energy. However, material recovery remains relatively constant, the amount of garbage dumped in landfills is substantially reduced by utilizing waste to generate electricity, hence lowering environmental pollution (Statistics Finland 2019). Norway will defeat Finland and take first position in the area of the planet and climate in 2021. (www.goodcountry.org).

6.3 Importance of Liquid Biofuels

Consumers have recently become more interested in liquid biofuels, which could be an effective method to deal with the exhaustion of fossil fuels. Liquid fuels are easier and safer to store and carry than gas fuels, and they have larger calorific values per volume. Also, have attributes close to diesel, gasoline, and other petroleum-based fuels, and they have the potential to reduce GHG emissions in the transportation sector by 70–90% with only minor changes to technology of vehicles and current infrastructure of fuel distribution [19]. Because of severe disposal principles, resource shortages, global emission consequences, and general environmental health and protection concerns, reduction of waste, recycling, and recovery of useful materials have arisen as highly significant matters. The current municipal trash management strategy reflects this philosophy. Food waste, domestic garbage, and food waste from, hotels, cafeterias, lunchrooms, and markets, as well as orchard waste, make up the organic fraction of municipal solid waste (OFMSW) [20, 21]. Due to its high availability and low cost, energy from organic fraction of MSW could be a technically and economically feasible alternate [22, 23]. Food waste makes up the majority of OFMSW. In the EU, approximately, 130 million tn of organic waste is generated each year, with food waste accounting for 69% [24, 25]. In China, the rate of municipal solid waste (MSW) production is roughly 170 million tn/y, with food waste accounting for 55% [26, 27]. Similarly, despite MSW creation being smaller (50 million tn/y) in Indian cities, 40–60% of MSW is considered organic [28]. Irrespective of the area food generation rate as an absolute number, over 35% of edible food is wasted [29].

6.4 Production of Bioethanol from Organic Fraction of MSW

The study of second-generation bioethanol has been ongoing for decades [14, 30]. A stage-by-stage method is used to explain the production of bioethanol from OFMSW in order to clarify the published findings. Pretreatment of feedstock, enzymatic saccharification, fermentation, product recovery, and then waste treatment are among the stages of the process.

6.4.1 Pretreatment

The pretreatment aims to change and disrupt structural properties of OFMSW in order to facilitate enzyme access and increase sugar monomer synthesis [31].

The initial step for production of bioethanol from OFMSW can be achieved through grinding, chipping, and milling [32]. Mechanical pretreatment can also be improved through the biodegradability of organic fraction of MSW. This is a key process in any kind of pretreatment method [33]. The hydrothermal pretreatment seeks to change the insoluble fraction's structural properties, making it more biodegradable [34].

As the temperature raised from 80 to 100 °C, hydrothermal pretreatment with hot dist. water lead to 20% enhancement in production of glucose [35–37], whereas pretreated samples with lesser temp., i.e., hot water (60 °C) had somewhat lower glucose concentrations (22.7 g/L) than the control samples (24.7 g/L) [38]. Hydrothermal treatment, in addition to being a pretreatment method, has been proposed as a sorting strategy for MSW [39]. Waste is treated under thermal degradation in a steam environment (160 °C); this process is also known as "active hygienization," and it produces diverse results for example organic fiber, glass, metal, plastic, and other components [39].

If there is any phenolic matrix in OFMSW, alkaline pretreatment digests it and renders cellulose and hemicellulose accessible for enzymatic breakdown [40]. Kerimakner et al. [41] found that chemical pretreatment with 3% NaOH (90 min, 30 °C) enhanced soluble sugar content by 35%, however, Vavouraki et al. [42] found that alkaline pretreatment (NaOH 0.7%, 1.5%, 3%) at 75 °C reduced soluble sugars.

One of the most often used ways of increasing yield of sugars from MSW is acid pretreatment. Acid hydrolysis can be used as a pretreatment as well as a method of hydrolysis. In comparison to the untreated, Vavouraki et al. [42] found that employing acid pretreatment of OFMSW at 100 °C with 1.17% HCl (86 min) and 1.12% HCl (94 min) resulted in a considerable increase (120%) of soluble sugars. Kerimakner et al. [41] found that acid pretreatment with 1% HCl (90 min, 60 degrees Celsius) resulted in 95% higher sugar concentrations than the untreated waste. Alamanou et al. [37] also found that adding 1% H_2SO_4 to enzymatic hydrolysis increased C6-sugar solubilization by about 7 and 13% at 85 and 100 degrees Celsius, respectively. The rate of hydrolysis was unaffected by H_2SO_4 concentrations greater than 1% [43]. H_2SO_4 with steam (121 °C, 134 °C, 15 min) and H_2SO_4 with microwave (700 W, 2 min) combined treatments yielded higher glucose yields than H_2SO_4 alone [43].

6.4.2 Enzymatic Hydrolysis

Enzymatic hydrolysis is a critical step in the generation of bioethanol from OFMSW. The target components are the structural carbohydrates cellulose, hemicellulose, and starch, which release pentoses and hexoses during hydrolysis, which can then be used in the fermentation process. Starch has been identified as the main target molecule for enzymatic saccharification in several studies. Glucoamylase Nagase N-40 outperformed Glucochimu #20,000, according to Tang et al. [44]. When enzymatically hydrolyzing wet SS-OFMSW (20% TS) with a-amylase 10U/g and glucoamylase 142.2U/g at 55 °C for 2.48 h, Yan et al. [45] obtained 99% saccharification yield. The hydrolysate had a lowering sugar content of 98.75 g/L (85%) at comparable circumstances but with a lower solid loading (15%) [46]. When enzymatic hydrolysis was achieved at 60 °C for 10 h and 8% solid loading with glucoamylase (85U/mL), Hafid et al. [47] showed a reduced degree of saccharification (59.90%). Nonetheless, prior to enzymatic hydrolysis, an acid pretreatment (1.5% HCl, 90 °C, 3 h) enhanced the saccharification yield to 86.4%.

6.4.3 Fermentation

Bioethanol is the metabolic result of fermentation of OFMSW hydrolysate. Continuous, fed-batch, and batch fermentation systems are the main three types of fermentation systems. The properties of the hydrolysate, yeast kinetic parameters, and techno-economical properties all play a role in determining which system is best [48]. For ethanol recovery by distillation, a minimum concentration of ethanol (4% w/v) should be considered a viability threshold [49]. High solids loadings could ensure a high ethanol concentration. It was technically difficult to achieve initial OFMS concentrations of more than 10–20%. In CSTRs, the initial viscosity is quite high, making stirring difficult, as well as causing heat and mass transfer issues and increased power consumption [50]. The top handleable limit in pilot-scale applications has been defined as 15–20% solid loading [51]. Fermentation can occur after or simultaneously with hydrolysis, depending on the operational mode [33]. As a result, a number of different fermentation process setups have been proposed. Bioethanol production from organic fraction of MSW has been reported using separate hydrolysis and fermentation (SHF), simultaneous saccharification and fermentation (SSF), non-isothermal simultaneous saccharification and fermentation (NSSF), simultaneous saccharification and co-fermentation (SSCF), and continuous immobilized fermentation (Table 6.1).

Table 6.1 Bioethanol production from OFMSW through fermentation

Feedstock	Enzymes	Organism	Fermentation	Yield of Ethanol	Reference
Hydrothermally pretreated OFMSW	Liquezyme, CellicCTec2, HTec2	*Mucor indicus CCUG 22,424*	SHF	52.5	53
Acidic pretreatment of OFMSW	*Aspergillus niger*, HTec2	*Mucor indicus*	SHF	53.3	54
SS-OFMSW	Alpha-amylase, glucoamylase	*Saccharomyces cerevisiae KCTC7107*	SHF	85.38	36
LAB sprayed SS-OFMSW	Nagase N-40 glucoamylase	*Flocculating Saccharomyces cerevisiae KF-7*	SHF	78.6	55
Thermally pretreated OFMSW	Cellulase (NS50013), Beta-glucosidase (NS50010)	*Saccharomyces cerevisiae*	SSF	0.6	46

Microorganism for fermentation

Saccharomyces cerevisiae is the most widely used microbial strain for ethanol synthesis [33], and it has a high tolerance for inhibitory chemicals [54]. Its impact on OFMSW has been thoroughly researched [35, 45]. It has also been found that the yeast does not require any additional nutrients during OFMSW fermentation [37, 38]. Due to the higher survival rate of organism and the high viable cell concentration in the fermenter resulting from its great flocculating properties, using the flocculating yeast, i.e., *Saccharomyces cerevisiae* strain KF-7 for fermentation of organic fraction of municipal solid waste could result in higher ethanol production and yield [44]. *Saccharomyces cerevisiae, Candida parasilopsis, and Lachancea fermentati* were isolated from naturally fermented nypa sap and successfully fermented sugars contained in the OFMSW hydrolysate for the generation of bioethanol [47]. Cell immobilization has been presented as a way to increase bioethanol production yields while lowering energy demands. The immobilization of Saccharomyces cerevisiae H058 for constant ethanol fermentation of OFMSW has been explored using calcium alginate gel beads [46]. Other bacteria have been identified as ethanologens in addition to Saccharomyces cerevisiae. In comparison to normal yeasts, *Zymomonas mobilis*, an acid-tolerant ethanol strain, produced more ethanol during OFMSW fermentation. [55, 56] Due to Saccharomyces cerevisiae's inability to ferment C5-sugars, researchers are looking for strains that can use pentose [54] to fully exploit the carbohydrate content of OFMSW. Mahmoodi et al. [52, 53] employed *Mucor indicus CCUG 22,424*, a Zygomycetes fungus strain, to perform ethanolic fermentation. Despite having a higher inhibitor concentration than baker's yeast, it was able to ferment both hexose and pentose carbohydrates. Ntaikou et al. [57] used co-cultures

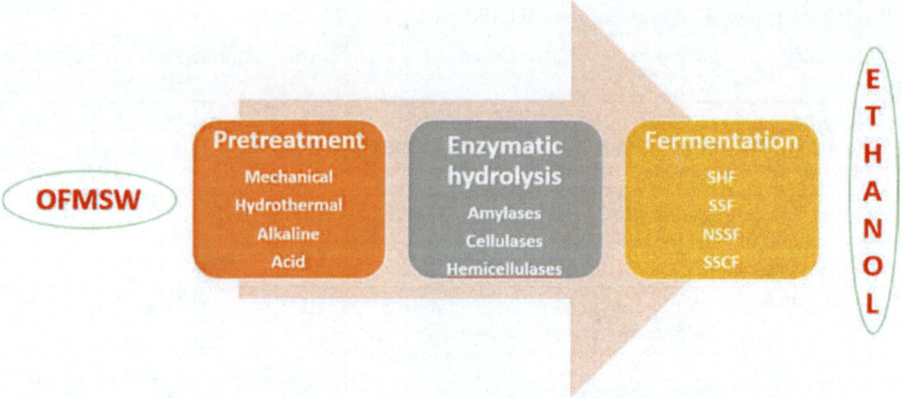

Fig. 6.2 Schematic representation of ethanol production from OFMSW

of *Saccharomyces cerevisiae CECT 1332* and *Pichia stipitis CECT 1922* to investigate the conversion of OFMSW to ethanol. Both were able to ferment pentoses and hexoses at the same time (Fig. 6.2).

6.4.4 Recovery of Ethanol

When anhydrous ethanol is the desired output, bioethanol can be recovered from the fermentation broth on a large-scale using distillation and molecular sieves. Nonetheless, because ethanol inhibits the fermenting microorganisms, concentration of bioethanol in the fermentation broth stays low. In situ bioethanol removal is an effective technique to overcome this barrier, resulting in higher production rate and lower process costs [54].

6.5 Role of Circular Economy

Separating, sorting, marketing, and preparing materials removed from the solid waste stream, as well as changing or remanufacturing those resources for use as feedstock for new things and other gainful employments, are all examples of reusing. Circular economy is achieved by the 3R system (Resource, Reuse, and Recovery), which results in economic enhancement through resource sustainability. CE has recently received a lot of attention in the field of industrial economics, with an emphasis on waste prevention, resource efficiency, and generating employment [58, 59]. With the belief in this circular economy, including waste management, reuse, and recycling, MSW has huge potential for the generation of renewable energy [60]. By developing integrated and multifunctional forms for the utilization of biomass/waste in the creation of appealing intermediates and final products, biorefinery ideas could be a critical component for the progression of the circular economy. A critical component of the design of such bioprocesses is monetary feasibility, which corresponds to natural supportability, as seen by a decreased carbon footprint [61]. The organic fraction of municipal solid waste (OFMSW) could be an effective feedstock for anaerobic digestion and a promising source of biogas.

6.6 Techno-economic Challenges in the Production of Fuel from OFMSW

The valorization of organic fraction of municipal solid waste via bioethanol and biodiesel routes has been the subject of extensive research. Nonetheless, each treatment train faces some technological difficulties. Enzyme cost, heat generation during process, and energy required for distillation have all been recognized as key barriers in the manufacturing of bioethanol. On the other hand, various issues connected to the nature of the catalyst (acidic, alkaline) still exist in the manufacture of biodiesel, posing economic and environmental constraints. The requirement of lower water content attained by normal drying processes or lyophilization for effective oil liberation adds to the cost of biodiesel manufacturing. Another barrier to fossil fuel replacement is the variability in oil recovery as well as the deprived marketability of the glycerol produced. Despite the fact that these technical hurdles are solvable, the economic sustainability of each option is questionable. The latter can be handled if both ethanol and biodiesel production are combined within a biorefinery that valorizes the produced residues at the same time. This could increase environmental performance while also lowering the cost of the process [14].

6.7 Utilization of MSW for Commercial Scale

The emergence of a long-term OFMSW valorization sector necessitates a complex supply chain that includes separate collection and transport of organic fraction of MSW, its processing, and the release of end-products to the market. Nonetheless, there are other issues that must be addressed. Because OFMSW collecting is difficult due to its disorganized nature, efforts should be made to make it much faster and more efficient. Regional and seasonal variations in the composition of OFMSW exist. As a result, this characteristic should be considered in the scale-up design, assuring a consistent and stable supply of feedstock. Purification of end-products is another difficulty that should be addressed during scale-up. The interoperability of biofuel products with the fossil fuel delivery network is critical. It should also be guaranteed that its quality fulfills the product conditions and standards. Only by incorporating good co-product use into integrated biorefineries will they be able to operate at a cost-effective level. For the integrated process to be commercialized successfully, each stage along the value chain from MSW to added-value goods must have favorable economics. Policymakers and researchers have emphasized the utilization of non-edible and low-cost residual waste for both bioenergy and fuel generation, putting an end to the "fuel vs. food" dispute. To effectively address these difficulties, national programs should support waste management and green methods. Obligations to use biofuels and tax cuts could help in this approach. While exploring new commercial frontiers, oil firms should also push fossil fuel alternatives. Consumer education would expose the benefits of bioproducts, resulting in a shift in consumer behavior and widespread adoption. To make biofuel production more cost-effective, the scientific and corporate communities, as well as for deciding members, should form a joint scheme to convert OFMSW to bioenergy. OFMSW biorefineries could be realized with the right policy and financial assistance [14].

6.8 Impact on Environmental Aspect

The demand for environmentally sound MSW handling has risen dramatically in recent decades. Recovering energy from garbage is a growing trend in most countries of the world today. If no changes are made in the area, annual CO_2-equivalent will exceed 2.6 billion tons by 2050 [6]. The organic part of the disintegrated waste releases CH_4, which is nearly 20 times more capable than CO_2 in terms of global emission potential [8]. Landfills contribute roughly 30% of global methane emissions. Composting is the ideal approach since it reduces greenhouse gas emissions and the amount of leachate produced by landfilling. When composting is incorrectly handled and done, it can result in environmental concerns such as the development of poisonous gases, aerosols, and dust, as well as work-related health problems [62]. According to a Behrooznia comparative life cycle evaluation, the anaerobic digestion-based MSW management technology reduced human health damage by 66.67%, environmental damage by 47.84%, and climate change by 89.64% in the Rasht City of Iran [63]. The conversion of both liquid and solid waste to power and bioenergy at the same time reduces harmful chemicals and increases biogas generation [64]. Cudjoe and Han investigated the biogas-derived production of power utilizing MSW organic compounds as a source, which were collected from 31 provinces in China and treated in landfills and AD technologies between 2004 and 2018. The findings revealed that landfill gas to bioenergy technology lowers down the global emission by 71.5%, while anaerobic digestion cuts it by 92.7% [65]. The energy and resource recovery potential of AD technology is unmatched by other waste treatment technologies. LCA researchers were asked to solve severe difficulties, such as the environmental impact valuation of the unorganized waste industry or informal waste sector [66]. Annually, fossil fuel burns will produce around 21.3 billion tons of CO_2, as well as a large number of hazardous gases such as nitrogen oxides, carbon monoxide, volatile organic compounds (VOCs), and particulate matter fractions. Researchers observed that the incineration process produced energy equivalent to around 15%

of the trash generated, which is said to be three times more beneficial than typical biogas production by landfills [67]. When compared to direct transfer to a clean landfill location, Taskin and Demir's study found that building a transfer section for MSW reduced global warming potential (GWP) impacts by 44.9% and CED scores by 51.7% [68]. Gasification can be a potential performance technology for MSW treatment because it tackles the environmental concerns better by converting it to energy [69]. According to a study, plasma gasification of 1 ton MSW has a global warming potential of 31 kg CO_2 eq. Overall, pyrolysis/gasification-based waste to energy conversion must be developed as an entire process chain to be commercially successful [70]. The environmental implications of the aforesaid procedures were listed in the following order: landfill > pyrolysis > incineration > gasification, based on the above studies. Any product, regardless of technique, MSW has an environmentally favorable characteristic.

6.9 Economics of Conversion from MSW to Energy

MSW is one of the difficulties that has arisen as a result of the economic and urban population increase. A P-graph was used to examine the potential turnover of 42 €/t of treated MSW when applying CE in waste management schemes [71]. In South Africa, scientists evaluated the economic viability of hybrid-derived anaerobic assimilation and gasification of urban garbage for power generation and determined that waste-to-energy is feasible [72]. *S. cerevisiae* and *Z. mobilis* are more economically viable than *E. coli* and *S. pombe* for producing bioethanol from municipal solid waste. Based on several aspects like autoclave pretreatment, reduced enzyme consumption, and increased solid loading, viable ethanol production from MSW can be achieved [73]. A study has optimized pectin, ethanol, and biogas production. From 1 t of orange waste, the highest yield was 244 kg of pectin, 26.5 L of ethanol, and 36 m^3 of CH_4 [74]. The integrated MSW management facility is both environmentally and financially supportable, with secondary product replacements offsetting environmental and financial costs (e.g., recycled plastics, metals, and bricks). In terms of cost, replacing natural gas, coal gas, and electricity with biogas has the potential to save 11.8%, 18.9%, and 8.5% of the overall cost, respectively, as compared to coal replacement [75].

6.10 Future Recommendations

The majority of experimental results are only applicable to small-scale production, but industrial expansion necessitates a system of research or development sectors focused on yield improvement. This idea could prove beneficial to most densely occupied and developing nations, helping them to improve their economic standing. The ultimate goal of research experts should be to reduce the number of processing steps by integrating or assimilating a few of them. This could lower the cost of process maintenance while also reducing the amount of time it takes. Furthermore, more advanced procedures should be used to improve production output. The methods carried out must be more useful so that the consumed substrate can be used either for secondary manufacturing of other products or as an alternate source of nutrition for different crops rather than being discarded. Only when their applicability is recognized in the energy-reliable industry will biofuel generation from MSW reach incredible heights. Nations' ruling authorities must treat this as a global issue and enact legislation making the judiciary use of alternative biofuels derived from renewable resources required in energy-intensive industries. Wind energy is extensively employed for power generation, and the requirements of biodiesels, biobutanol, biogas, and other biofuels should be acknowledged internationally, similar to how solar cells (source of solar energy) have substituted the call for heaters [76]. Apart from biofuel production, MSW clearing has some aesthetic benefits, such as the presence of clear air, less environmental degradation, and the potential to become a tourist attraction. Tax cuts and government assistance aid in the development of the profitable WtE plant. When compared to isolated processes, integrated processes are more effective and lucrative. The development of WtE plants minimizes global warming potential as well as energy demand.

6.11 Summary

This chapter outlines OFMSW's inherent potential for resource recovery within a sustainable agenda. Furthermore, the primary mechanisms of organic fraction of MSW that can be valorized through biofuels production are highlighted. In this work, several technological options for OFMSW valorization were provided, including chemical and different enzymatic methods, as well as fermentation procedures employing appropriate strains. Several features of bioethanol generation from OFMSW are also covered in this literature study. Diverse range of aspects is discussed ranging from technical challenges in the way of production to economic considerations. As a result, it was demonstrated that OFMSW may be used as an appropriate feedstock for the manufacture of high-quality ethanol, potentially contributing significantly to the bio-based economy.

References

1. Usmani, Z., Kumar, V., Varjani, S., Gupta, P., Rani, R., & Chandra, A. (2020). Municipal solid waste to clean energy system: a contribution toward sustainable development. In *Current Developments in Biotechnology and Bioengineering* (pp. 217–231). Elsevier.
2. Pavi, S., Kramer, L. E., Gomes, L. P., & Miranda, L. A. S. (2017). Biogas production from co-digestion of organic fraction of municipal solid waste and fruit and vegetable waste. *Bioresource Technology, 228*, 362–367.
3. Tang, Z., Li, W., Tam, V. W., & Xue, C. (2020). Advanced progress in recycling municipal and construction solid wastes for manufacturing sustainable construction materials. *Resources, Conservation & Recycling: X, 6*, 100036.
4. Istrate, I. R., Iribarren, D., Gálvez-Martos, J. L., & Dufour, J. (2020). Review of life-cycle environmental consequences of waste-to-energy solutions on the municipal solid waste management system. *Resources, Conservation and Recycling, 157*, 104778.
5. Neehaul, N., Jeetah, P., & Deenapanray, P. (2020). Energy recovery from municipal solid waste in Mauritius: Opportunities and challenges. *Environmental Development, 33*, 100489.
6. Yadav, V., & Karmakar, S. (2020). Sustainable collection and transportation of municipal solid waste in urban centers. *Sustainable Cities and Society, 53*, 101937.
7. Rajmohan, K. S., Chandrasekaran, R., & Varjani, S. (2020). A review on occurrence of pesticides in environment and current technologies for their remediation and management. *Indian Journal of Microbiology, 60*(2), 125–138.
8. Yaman, C., Anil, I., & Alagha, O. (2020). Potential for greenhouse gas reduction and energy recovery from MSW through different waste management technologies. *Journal of Cleaner Production, 264*, 121432.
9. Louati, A. (2016). Modeling municipal solid waste collection: A generalized vehicle routing model with multiple transfer stations, gather sites and inhomogeneous vehicles in time windows. *Waste Management, 52*, 34–49.
10. Nikku, M., Deb, A., Sermyagina, E., & Puro, L. (2019). Reactivity characterization of municipal solid waste and biomass. *Fuel, 254*, 115690.
11. Kaza, S., Yao, L., Bhada-Tata, P., & Van Woerden, F. (2018). *What a waste 2.0: a global snapshot of solid waste management to 2050*. World Bank Publications.
12. Byun, J., & Han, J. (2021). Environmental analysis of bioethanol production strategies from corn stover via enzymatic and nonenzymatic sugar production. *Bioresource Technology, 328*, 124808.
13. Kwon, O., & Han, J. (2021). Organic-waste-derived butyric acid-to-biodiesel supply-chain network: Strategic planning design using a deterministic snapshot model. *Journal of Environmental Management, 293*, 112848.
14. Barampouti, E. M., Mai, S., Malamis, D., Moustakas, K., & Loizidou, M. (2019). Liquid biofuels from the organic fraction of municipal solid waste: A review. *Renewable and Sustainable Energy Reviews, 110*, 298–314.
15. Barik, S., & Paul, K. K. (2017). Potential reuse of kitchen food waste. *Journal of Environmental Chemical Engineering, 5*(1), 196–204.
16. Rane, N. M., Admane, S. V., Sapkal, R. S. (2019). Adsorption of hexavalent chromium from wastewater by using sweetlime and lemon peel powder by batch studies. In *Waste Management and Resource Efficiency* (pp. 1207–1220). Springer, Singapore.
17. Nizami, A. S., Rehan, M., Waqas, M., Naqvi, M., Ouda, O. K., Shahzad, K., Miandad, R., Khan, M. Z., Syamsiro, M., Ismail, I. M., & Pant, D. (2017). Waste biorefineries: Enabling circular economies in developing countries. *Bioresource Technology, 241*, 1101–1117.
18. Bhatia, S. K., Joo, H. S., & Yang, Y. H. (2018). Biowaste-to-bioenergy using biological methods–a mini-review. *Energy Conversion and Management, 177*, 640–660.
19. Zhang, Z., O'Hara, I. M., Mundree, S., Gao, B., Ball, A. S., Zhu, N., Bai, Z., & Jin, B. (2016). Biofuels from food processing wastes. *Current Opinion in Biotechnology, 38*, 97–105.
20. Campuzano, R., & González-Martínez, S. (2016). Characteristics of the organic fraction of municipal solid waste and methane production: A review. *Waste Management, 54*, 3–12.

21. Al Seadi, T., Owen, N. E., Hellström, H., & Kang, H. (2013). *Source Separation of MSW*.IEA Bioenergy.
22. Tyagi, V. K., Fdez-Güelfo, L. A., Zhou, Y., Álvarez-Gallego, C. J., Garcia, L. R., & Ng, W. J. (2018). Anaerobic co-digestion of organic fraction of municipal solid waste (OFMSW): Progress and challenges. *Renewable and Sustainable Energy Reviews, 93*, 380–399.
23. Romero-Cedillo, L., Poggi-Varaldo, H. M., Ponce-Noyola, T., Ríos-Leal, E., Ramos-Valdivia, A. C., Cerda-García Rojas, C. M., & Tapia-Ramírez, J. (2017). A review of the potential of pretreated solids to improve gas biofuels production in the context of an OFMSW biorefinery. *Journal of Chemical Technology & Biotechnology, 92*(5), 937–958.
24. Clarke, W. P. (2018). The uptake of anaerobic digestion for the organic fraction of municipal solid waste–push versus pull factors. *Bioresource technology, 249*, 1040–1043.
25. Seo, J. Y., Heo, J. S., Kim, T. H., Joo, W. H., & Crohn, D. M. (2004). Effect of vermiculite addition on compost produced from Korean food wastes. *Waste Management, 24*(10), 981–987.
26. De Clercq, D., Wen, Z., Fan, F., & Caicedo, L. (2016). Biomethane production potential from restaurant food waste in megacities and project level-bottlenecks: A case study in Beijing. *Renewable and Sustainable Energy Reviews, 59*, 1676–1685.
27. Zhou, H., Meng, A., Long, Y., Li, Q., & Zhang, Y. (2014). An overview of characteristics of municipal solid waste fuel in China: Physical, chemical composition and heating value. *Renewable and sustainable energy reviews, 36*, 107–122.
28. Sharholy, M., Ahmad, K., Mahmood, G., & Trivedi, R. C. (2008). Municipal solid waste management in Indian cities—A review. *Waste Management, 28*(2), 459–467.
29. Chatterjee, B., & Mazumder, D. (2016). Anaerobic digestion for the stabilization of the organic fraction of municipal solid waste: A review. *Environmental Reviews, 24*(4), 426–459.
30. Ho, D. P., Ngo, H. H., & Guo, W. (2014). A mini review on renewable sources for biofuel. *Bioresource Technology, 169*, 742–749.
31. Hafid, H. S., Shah, U. K. M., Baharuddin, A. S., & Ariff, A. B. (2017). Feasibility of using kitchen waste as future substrate for bioethanol production: A review. *Renewable and Sustainable Energy Reviews, 74*, 671–686.
32. Sarkar, N., Ghosh, S. K., Bannerjee, S., & Aikat, K. (2012). Bioethanol production from agricultural wastes: An overview. *Renewable Energy, 37*(1), 19–27.
33. John, I., Muthukumar, K., & Arunagiri, A. (2017). A review on the potential of citrus waste for D-Limonene, pectin, and bioethanol production. *International Journal of Green Energy, 14*(7), 599–612.
34. Wan, C., & Li, Y. (2011). Effect of hot water extraction and liquid hot water pretreatment on the fungal degradation of biomass feedstocks. *Bioresource Technology, 102*(20), 9788–9793.
35. Hafid, H. S., Nor'Aini, A. R., Mokhtar, M. N., Talib, A. T., Baharuddin, A. S., & Kalsom, M. S. U. (2017). Over production of fermentable sugar for bioethanol production from carbohydrate-rich Malaysian food waste via sequential acid-enzymatic hydrolysis pretreatment. *Waste Management, 67*, 95–105.
36. Hafid, H. S., Shah, U. K. M., & Baharudin, A. S. (2015). Enhanced fermentable sugar production from kitchen waste using various pretreatments. *Journal of Environmental Management, 156*, 290–298.
37. Alamanou, D. G., Malamis, D., Mamma, D., & Kekos, D. (2015). Bioethanol from dried household food waste applying non-isothermal simultaneous saccharification and fermentation at high substrate concentration. *Waste and Biomass Valorization, 6*(3), 353–361.
38. Cekmecelioglu, D., & Uncu, O. N. (2013). Kinetic modeling of enzymatic hydrolysis of pretreated kitchen wastes for enhancing bioethanol production. *Waste Management, 33*(3), 735–739.
39. Ballesteros, M., Sáez, F., Ballesteros, I., Manzanares, P., Negro, M. J., Martínez, J. M., Castañeda, R., & Oliva Dominguez, J. M. (2010). Ethanol production from the organic fraction obtained after thermal pretreatment of municipal solid waste. *Applied Biochemistry and Biotechnology, 161*(1), 423–431.
40. Pandey, A., Soccol, C.R., Nigam, P., & Soccol, V. T. (2000). Biotechnological potential of agro-industrial residues. I: sugarcane bagasse. *Bioresource Technology, 74*(1), 69–80.

41. Ahmed, B., Tyagi, V. K., Aboudi, K., Naseem, A., Álvarez-Gallego, C. J., Fernández-Güelfo, L. A., Kazmi, A. A., & Romero-García, L. I. (2021). Thermally enhanced solubilization and anaerobic digestion of organic fraction of municipal solid waste. *Chemosphere, 282*, 131136.
42. Vavouraki, A. I., Angelis, E. M., & Kornaros, M. (2013). Optimization of thermo-chemical hydrolysis of kitchen wastes. *Waste Management, 33*(3), 740–745.
43. Li, A., Antizar-Ladislao, B., & Khraisheh, M. (2007). Bioconversion of municipal solid waste to glucose for bio-ethanol production. *Bioprocess and Biosystems Engineering, 30*(3), 189–196.
44. Tang, Y. Q., Koike, Y., Liu, K., An, M. Z., Morimura, S., Wu, X. L., & Kida, K. (2008). Ethanol production from kitchen waste using the flocculating yeast Saccharomyces cerevisiae strain KF-7. *Biomass and Bioenergy, 32*(11), 1037–1045.
45. Yan, S., Li, J., Chen, X., Wu, J., Wang, P., Ye, J., & Yao, J. (2011). Enzymatical hydrolysis of food waste and ethanol production from the hydrolysate. *Renewable Energy, 36*(4), 1259–1265.
46. Yan, S., Wang, P., Zhai, Z., & Yao, J. (2011). Fuel ethanol production from concentrated food waste hydrolysates in immobilized cell reactors by Saccharomyces cerevisiae H058. *Journal of Chemical Technology & Biotechnology, 86*(5), 731–738.
47. Hafid, H. S., Abdul Rahman, N. A., Md Shah, U. K., Samsu Baharudin, A., & Zakaria, R. (2016). Direct utilization of kitchen waste for bioethanol production by separate hydrolysis and fermentation (SHF) using locally isolated yeast. *International Journal of Green Energy, 13*(3), 248–259.
48. Ch, A. K., Chan, E. S., Rudravaram, R., Narasu, M. L., Rao, L. V., & Ravindra, P. (2007). Economics and environmental impact of bioethanol production technologies: An appraisal. *Biotechnology and Molecular Biology Reviews, 2*(1), 14–32.
49. Wingren, A., Galbe, M., & Zacchi, G. (2003). Techno-economic evaluation of producing ethanol from softwood: Comparison of SSF and SHF and identification of bottlenecks. *Biotechnology Progress, 19*(4), 1109–1117.
50. Fan, Z., & Lynd, L. R. (2007). Conversion of paper sludge to ethanol. I: Impact of feeding frequency and mixing energy characterization. *Bioprocess and Biosystems engineering, 30*(1), 27–34.
51. Jørgensen, H., Kristensen, J. B., & Felby, C. (2007). Enzymatic conversion of lignocellulose into fermentable sugars: Challenges and opportunities. *Biofuels, Bioproducts and Biorefining, 1*(2), 119–134.
52. Mahmoodi, P., Karimi, K., & Taherzadeh, M. J. (2018). Hydrothermal processing as pretreatment for efficient production of ethanol and biogas from municipal solid waste. *Bioresource Technology, 261*, 166–175.
53. Bolzonella, D., Fatone, F., Pavan, P., & Cecchi, F. (2005). Anaerobic fermentation of organic municipal solid wastes for the production of soluble organic compounds. *Industrial & Engineering Chemistry Research, 44*(10), 3412–3418.
54. Banerjee, S., Mudliar, S., Sen, R., Giri, B., Satpute, D., Chakrabarti, T., & Pandey, R. A. (2010). Commercializing lignocellulosic bioethanol: Technology bottlenecks and possible remedies. *Biofuels, Bioproducts and Biorefining: Innovation for a sustainable economy, 4*(1), 77–93.
55. Ma, H., Wang, Q., Qian, D., Gong, L., & Zhang, W. (2009). The utilization of acid-tolerant bacteria on ethanol production from kitchen garbage. *Renewable Energy, 34*(6), 1466–1470.
56. Balat, M. (2011). Production of bioethanol from lignocellulosic materials via the biochemical pathway: A review. *Energy Conversion and Management, 52*(2), 858–875.
57. Ntaikou, I., Menis, N., Alexandropoulou, M., Antonopoulou, G., & Lyberatos, G. (2018). Valorization of kitchen biowaste for ethanol production via simultaneous saccharification and fermentation using co-cultures of the yeasts Saccharomyces cerevisiae and Pichia stipitis. *Bioresource Technology, 263*, 75–83.
58. Su, B., Heshmati, A., Geng, Y., & Yu, X. (2013). A review of the circular economy in China: Moving from rhetoric to implementation. *Journal of Cleaner Production, 42*, 215–227.
59. Geissdoerfer, M., Savaget, P., Bocken, N. M., & Hultink, E. J. (2017). The Circular Economy–A new sustainability paradigm? *Journal of Cleaner Production, 143*, 757–768.
60. Rathore, P., & Sarmah, S. P. (2020). Economic, environmental and social optimization of solid waste management in the context of circular economy. *Computers & Industrial Engineering, 145*, 106510.

61. Abad, V., Avila, R., Vicent, T., & Font, X. (2019). Promoting circular economy in the surroundings of an organic fraction of municipal solid waste anaerobic digestion treatment plant: Biogas production impact and economic factors. *Bioresource Technology, 283*, 10–17.
62. Chen, H., Tang, M., Yang, X., Tsang, Y. F., Wu, Y., Wang, D., & Zhou, Y. (2021). Polyamide 6 microplastics facilitate methane production during anaerobic digestion of waste activated sludge. *Chemical Engineering Journal, 408*, 127251.
63. Behrooznia, L., Sharifi, M., & Hosseinzadeh-Bandbafha, H. (2020). Comparative life cycle environmental impacts of two scenarios for managing an organic fraction of municipal solid waste in Rasht-Iran. *Journal of Cleaner Production, 268*, 122217.
64. Carlini, M., Mosconi, E. M., Castellucci, S., Villarini, M., & Colantoni, A. (2017). An economical evaluation of anaerobic digestion plants fed with organic agro-industrial waste. *Energies, 10*(8), 1165.
65. Cudjoe, D., & Han, M. S. (2020). Economic and environmental assessment of landfill gas electricity generation in urban districts of Beijing municipality. *Sustainable Production and Consumption, 23*, 128–137.
66. Awasthi, M. K., Sarsaiya, S., Wainaina, S., Rajendran, K., Awasthi, S. K., Liu, T., Duan, Y., Jain, A., Sindhu, R., Binod, P., & Pandey, A. (2021). Techno-economics and life-cycle assessment of biological and thermochemical treatment of bio-waste. *Renewable and Sustainable Energy Reviews, 144*, 110837.
67. dos Santos, I. F. S., Mensah, J. H. R., Gonçalves, A. T. T., & Barros, R. M. (2020). Incineration of municipal solid waste in Brazil: An analysis of the economically viable energy potential. *Renewable Energy, 149*, 1386–1394.
68. Taşkın, A., & Demir, N. (2020). Life cycle environmental and energy impact assessment of sustainable urban municipal solid waste collection and transportation strategies. *Sustainable Cities and Society, 61*, 102339.
69. Chanthakett, A., Arif, M. T., Khan, M. M. K., & Oo, A. M. (2021). Performance assessment of gasification reactors for sustainable management of municipal solid waste. *Journal of Environmental Management, 291*, 112661.
70. Ramos, A., Berzosa, J., Espi, J., Clarens, F., & Rouboa, A. (2020). Life cycle costing for plasma gasification of municipal solid waste: A socio-economic approach. *Energy Conversion and Management, 209*, 112508.
71. Van Fan, Y., Klemeš, J. J., Walmsley, T. G., & Bertók, B. (2020). Implementing Circular Economy in municipal solid waste treatment system using P-graph. *Science of The Total Environment, 701*, 134652.
72. Mabalane, P. N., Oboirien, B. O., Sadiku, E. R., & Masukume, M. (2021). A techno-economic analysis of anaerobic digestion and gasification hybrid system: Energy recovery from municipal solid waste in South Africa. *Waste and Biomass Valorization, 12*(3), 1167–1184.
73. Meng, F., Dornau, A., Mason, S. J. M., Thomas, G. H., Conradie, A., & McKechnie, J. (2021). Bioethanol from autoclaved municipal solid waste: Assessment of environmental and financial viability under policy contexts. *Applied Energy, 298*, 117118.
74. Vaez, S., Karimi, K., Mirmohamadsadeghi, S., & Jeihanipour, A. (2021). An optimal biorefinery development for pectin and biofuels production from orange wastes without enzyme consumption. *Process Safety and Environmental Protection, 152*, 513–526.
75. Wang, Z., Lv, J., Gu, F., Yang, J., & Guo, J. (2020). Environmental and economic performance of an integrated municipal solid waste treatment: A Chinese case study. *Science of the Total Environment, 709*, 136096.
76. Rajendran, N., Gurunathan, B., Han, J., Krishna, S., Ananth, A., Venugopal, K., & Priyanka, R. S. (2021). Recent advances in valorization of organic municipal waste into energy using biorefinery approach, environment and economic analysis. *Bioresource Technology, 337*, 125498.

Chapter 7
Algae as Potential Feedstock for Bioethanol Production

Abstract Alternative renewable fuels are becoming increasingly important due to concerns about fuel security, the economy, and climate change. This study provides a thorough overview of algae harvesting and processing technologies as well as their applications in the development of biofuels like bioethanol. Because of their rapid growth, microalgae have been identified as a potential pollution control agent and a viable replacement for currently used non-renewable resources. Algal biofuels have the potential to be a renewable energy source. Algae are a good 3G feedstock for bioethanol production because of their fast development, high biomass yield, and high lipid and carbohydrate content. Algae are projected to overtake 1G and 2G feedstocks soon due to their enormous potential.

Keywords Algal Bioethanol · Cultivation · Harvesting · Pretreatment · Fermentation

7.1 Introduction

Microalgal biofuels are replacing traditional fossil fuels in the era of the replaceable energy revolution. Biomass for bioethanol production has been the subject of research for the past five decades. Generations of biofuels are classified. More crop-growing areas mean less space available for human and animal feed production, as well as massive-scale manufacturing of raw materials. Hence, the 1G biofuels debate has sparked a serious political and economic debate. Because the 1G feedstock is impractical, researchers are concentrating on the 2G biofuels, which are made from non-food materials such as cellulose biomass, inedible sections of crops, and various types of straw, manure, and wood. Due to the high expense of the production process, biofuel made from 2G feedstock sources is not commercially viable [1, 2]. As a result, scientists are focusing on the development of the 3rd era (3G) microalgae raw material, which is regarded as a sustainable energy source for biofuel manufacture in the current circumstances due to the decrease in both 1G and 2G ethanol sources [1, 3, 4]. Algae are simple organisms that use light for photosynthesis and contain chlorophyll. Algae are the quickest growing plants and are classified as macroalgae or microalgae based on their form. Macroalgae, also known

N. K. Aggarwal et al., *Bioethanol Production*, Green Chemistry and Sustainable Technology, https://doi.org/10.1007/978-3-031-05091-6_7

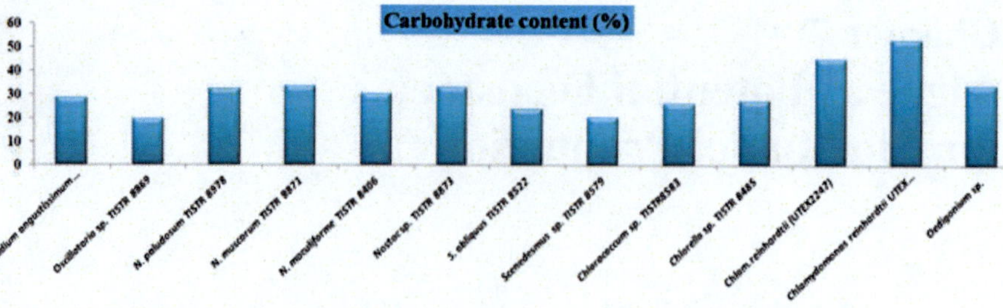

Fig. 7.1 Various carbohydrate concentrations of microalgae that are used to produce bioethanol

as "seaweeds," are multicellular oceanic plants that can grow to be 60 m long. In freshwater, seawater, and even wastewater, microalgae live in small cells of about 2–200 mm in diameter. Using sunlight, carbon dioxide, and nutrients from the water, they efficiently convert solar energy into biomass [5]. Algae biomass may be used to produce ethanol that is both sustainable and environmentally beneficial. Microalgae growing under normal conditions, that is, without food constraint, have a biochemical composition dominated by proteins (30–50%), carbohydrates (20–40%), and lipids (8–15%) [6, 7]. As far as algal bioethanol production goes, screening and selection are critical factors. There are over 100,000 different types of algae on the planet [8, 9]. Because of factors like rate of growth, photosynthetic yield, biomass productivity, sugar and fat content, and environmental adaptation, not all algae may be suitable for ethanol production. The ability to dominate wild strains in open pond production environments, high CO_2 sinking ability, limited nutrient requirements, tolerance to a wide temperature range resulting from the diurnal cycle and seasonal fluctuations, providing valuable co-products, and exhibiting self-flocculation character are just a few of the other advantages [2]. Since the 1970s, microalgae have been studied as a raw material for the production of biodiesel due to their widespread availability and higher oil yields than traditional terrestrial plants [10]. Research into the use of microalgae as an advanced energy material for ethanol production has seen a dramatic increase. For ethanol generation, species including *Chlorella, Dunaliella, Chlamydomonas, Scenedesmus*, and *Spirulina* are usually chosen [11]. Microalgae species' carbohydrate content is shown in Fig. 7.1 [11].

7.2 Algal Bioethanol

Clean and renewable, bioethanol can be produced by a simple microbial fermentation process and has many benefits over other biofuels. Brazil, Canada, and the United States all have commercial production of bioethanol, which is a promising green fuel alternative aimed at reducing carbon emissions in the transportation sector [5]. Because of their broad availability, high sugar content, and higher yield than other crops, algae are regarded as the best feedstock for bioethanol production. Algae are divided into two types based on their form and size: microalgae and macroalgae.

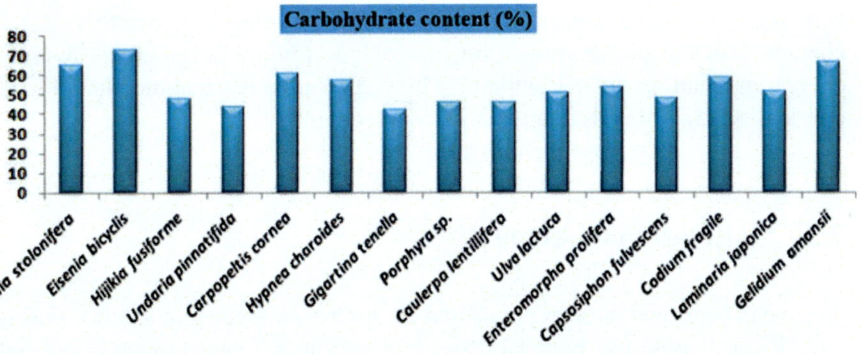

Fig. 7.2 Various carbohydrate concentrations of seaweeds that are used to produce bioethanol

Microalgae are minute photosynthetic organisms, many of which are unicellular, as their name suggests. Macroalgae, such as kelp, are made up of numerous cells that are arranged into structures that resemble the roots, stems, and leaves of higher plants [12]. The marine ecosystem is thought to generate almost half of the world's biomass [13]. In reality, algae that can grow in saline or municipal wastewater are essential for long-term bioethanol production since they do not compete with food crops that need freshwater for irrigation. Algae can also provide long-term wastewater bioremediation by using nitrogen and phosphorus from a variety of wastewater sources, such as agricultural run-off, concentrated livestock feed operations, and municipal and industrial wastewater [14, 15]. Macroalgae, or large algae, can be used for the production of bioethanol by converting their stored material to sugars, just like microalgae. Because lignin is absent or almost absent, algal cellulose is simple to hydrolyze using enzymes. Macroalgal genera such as *Laminaria, Saccorhiza,* and *Alaria* are brown algal genera that grow up to a meter in length and rely on laminarin and mannitol as their primary food sources [16–18]. Figure 7.2 depicts the carbohydrate content of different seaweeds [19].

Macroalgal bioethanol production has recently been investigated in the Netherlands, the United States, and Ireland. The International Energy Agency (IEA) has released a study on macroalgae as well as the US Department of Energy's roadmap for algal biofuels. Research on the use of macroalgae as a biofuel feedstock is underway in Asia and Europe [20]. Furthermore, due to their larger size, seaweeds are easier to cultivate and harvest and they may produce far more biomass than microalgae. As a result, the focus of current research has switched to the production of bioethanol from macroalgae. Due to two factors, seaweed is an underutilized potential source for bioethanol production. In the first place, the high biomass production of macroalgae is combined with the low capital and operating expenses of aquacultured marine farms or natural stocks of macroalgae. Compared to microalgal open ponds, these cultivation systems have lower startup and operating costs, but they produce more biomass. Seaweeds, in general, can survive a wide range of environmental circumstances. It is now possible to develop a suitable macroalgae species for bioethanol production because of advances in genetic engineering. Macroalgae that have been genetically modified would need to be grown in enclosed bioreactors [21]. These characteristics inspire optimism for macroalgae's future development in renewable energy applications such as ethanol. Microalgae are a feasible bioethanol feedstock due to their high biomass output and sugar content.

7.3 Cultivation of Algae

Algae can be grown in open ponds that are both man-made and natural. This is easy and cheap. People use both labs and ponds to grow algal biomass, and both are used. Microalgae need four things to grow and make food: light, CO_2, water, and organic fertilizers. To grow microalgae, there are many different ways to do it. These include closed-culture systems like photobioreactors (PBRs), as well as open ponds that include raceway ponds and hybrid systems [22, 23]. Algae have a high ability to take carbon dioxide out of the air, a high level of photosynthesis, and a high rate of growth. As a result, the number of nutrients in the wastewater decreases. They also use nitrogen and phosphorous from wastewater from municipalities, farmlands, and industries. They have a lot of lipids that can be used to make biofuels that are not harmful and that break down quickly. Algae cultivation for biofuel production appears simple because algae require few resources to grow. The amount of biomass, metabolism, and cell biochemical composition are all affected by pH levels in algae farming. Another thing that affects the growth of microalgae is the pH of the water. Species of microalgae have different pH needs [24]. The pH range of 6–8.76 is ideal for most of them. Most algae species' pH is sensitive, and only a few can sustain a pH range as wide as *C. vulgaris*', though pH 9–10 produces the highest growth rate and biomass production. A high pH will increase salinity, which is extremely harmful to algal cells [25].

7.3.1 Laboratory Cultivation

According to an early study, algae used for ethanol production may be cultivated in the lab using a range of growth media and under carefully controlled conditions. Algae beakers, plastic pots, and tubs are commonly used to raise algae. Under laboratory conditions, a variety of parameters such as artificial light duration and intensity, aeration, CO_2 concentration, temperature, pH of the growth medium, and nutrition delivery influence algal biomass output. To maintain optimal biomass production, these variables must be managed. In one study, temperature and pH were found to have an impact on bioethanol production, with S. cerevisiae IFST-072011 producing the highest output of bioethanol at a pH of 5.0–6.0 at 30 °C [26].

7.3.2 Raceway Ponds

Algae, water, and nutrients move in a racetrack-like pattern in these open ponds, which are referred to as "raceway ponds." For microalgal growth, open pond systems are the most common, and huge commercial systems of this type are currently in use. Since the 1950s, raceway ponds have been used to mass-produce microalgae. The pond is constructed of a 15–20 cm deep closed-loop recirculation canal with a paddlewheel. The broth can be retrieved from the same place where the culture is fed during the day [22]. The paddlewheel mixes circulate and prevent sedimentation of the algae biomass. These systems face a serious problem in that they are susceptible to contamination by bacteria or other microalgae [27]. An open raceway pond of 1800 L capacity was used to cultivate Botryococcus braunii Kutz for biofuel production and yielded a biomass concentration of 1.8 g/L [28].

7.3.3 Photobioreactor Systems (PBRs)

Algae bioreactors, or photobioreactors, are bioreactors used to grow algae in a controlled environment. Due to its ability to grow desirable species without altering their physical and biological properties, PBR is an ideal alternative to raceway ponds [4, 23]. According to Bajhaiya et al. [29], the tubular photobioreactor appears to be the best bioreactor for creating algal biomass. In a tubular PBR, fresh culture media is fed at a steady rate, and the same amount of microalgal broth is extracted at the same time every day. Tubular photobioreactors are constructed from plastic or glass tubes arranged in a grid pattern. The solar collector, which is a tube array, has a diameter of fewer than 0.1 m. Reducing the temperature at night can reduce the loss of biomass due to respiration. Heat exchangers can efficiently and cheaply cool outdoor tubular photobioreactors. The degassing column may have a heat exchange coil. The tube width is limited because light cannot penetrate too deeply into the dense culture broth necessary for high biomass yield in the photobioreactor [4].

7.3.4 Wastewater

Microalgae can be grown using wastewater and carbon dioxide as low-cost medium components. Algae biofuel production is economically and environmentally sustainable in several studies [30], with carbon nutrients accounting for roughly 60% of total output [31]. To resolve this issue, cheap and effective resources had to be found and exploited. Wastewater has been discovered to be a viable and cost-effective substrate for algae cultivation [30]. It has been discovered that microalgae can produce more biomass for biodiesel generation when cultivated in treated wastewater than when grown in traditional synthetic media. Similarly, if the water used to grow algae is deficient in nutrients, more nutrient sources will be necessary, raising production costs [32].

7.4 Harvesting of Algal Biomass

To produce bioethanol at a cheap cost, efficient and effective harvesting of algal biomass from the growing medium is required. The recovery and harvesting of algae have a variety of methods in place [33]. To be effective, an algal harvesting method should be able to harvest a product that has a high dry weight proportion and a low cost. One or more stages of solid–liquid partitioning are required for algal biomass harvesting, which is a thought-provoking step in the production process [34] and accounts for roughly 20–30% of total costs [35]. Biomass harvesting technologies include filtration, centrifugation, flocculation, and flotation (Table 7.1) [36].

7.5 Pretreatment and Saccharification of Algal Biomass

Most of the carbohydrates found in algae cells are confined within the cells. As a result, to liberate the carbohydrate from the algae, a pretreatment step is normally required. Algae is made up of carbohydrates that are turned into bioethanol by microbes like yeast and bacteria. Pretreatment, hydrolysis, and fermentation are the three most prevalent procedures in the production of bioethanol.

7.5.1 Pretreatment and Hydrolysis for Production of Sugar

Using a chemical pretreatment approach to liberate monosacchride from microalgae biomass has recently been reported by a number of researchers [7, 45]. Pretreatment and hydrolysis reactions can be carried out simultaneously in microalgae due to their simple cellular structure and lack of lignin, which necessitates relatively mild reaction conditions (to hydrolyze the complex carbohydrate molecules to simple sugars) [7]. Ho et al. [7] discovered that 96% of the hexose yield from Chlorella vulgaris biomass could be recovered with a 1% diluted sulfuric acid concentration, a reaction temperature of 121 °C, and a reaction period of 20 min. Aside from acids, alkali is effective in the carbohydrate breakdown of microalgae [46]. Harun et al. [46] employed 0.75 (w/v) sodium hydroxide at 120 °C for 30 min to get 0.35 g/g of biomass glucose production from *Chlorococcum infusionum*. Sodium hydroxide can break down the cell wall of microalgae, causing the cell wall to lose its strength and the carbohydrate to leak out. The carbohydrate was then further hydrolyzed by the presence of NaOH into simple fermentable sugar (Table 7.2).

Table 7.1 Summary of the theory, benefits, and drawbacks of various algal harvesting methods

Harvesting technique	Theory	Benefits	Drawbacks	References
Flocculation	Cells aggregate by utilizing a flocculant, which can be chemicals (ferric sulfate, ferric chloride, and ammonium sulfate), bioagents (chitosan), and microbes to increase their size (bacteria)	Saving time	Large occupation area. The cost of flocculants and operators is very high	[2, 35–38]
Filtration	Smaller cells (size 30 m) require ultrafilters to be gathered, whereas larger cells (size >70 m) can be filtered under pressure or suction. Membrane sheets with a ceramic coating can be used instead of conventional membranes	Saving time	Membrane fouling or clogging	[39]
Flotation	Bubbles of air are used to trap algae cells	It is less costly than other approaches	Depends on the distribution of bubbles in the suspension	[2]
Sonication	The use of acoustic forces to constantly pump organisms into a resonator chamber	No cell harming, no shear, small occupation space, continuous operation	The cooling system uses a large amount of power. Large-scale systems have a high unit cost	[40, 41]
Centrifugation	Cell size and density are factors in sedimentation	Rapid, Large microalgae can be grown in this system	Contains parts that can move independently	[2, 42, 43]
Precipitation	Self-precipitation occurs in some algae. After halting circulation, they sink to the bottom	It does not use any chemicals or energy	Species-specific. The length of time varies according to the species. The self-precipitation of some species is not common	[44]

Table 7.2 Production of ethanol from algal biomass using different hydrolysis methods

Hydrolysis technique	Microalgae	Carbohydrate (%)	Fermentable sugar (g/g algae)	Ethanol production (g/g algae)	Reference
Chemical					
Sodium hydroxide	*Chlorococcum infusionum*	32.5	0.35	0.26	[46]
Sulfuric acid	*Chlorella vulgaris*	51	0.5	0.233	[7]
Enzyme					
Endoglucanase, b-glucosidase and amylase	*Chlorella vulgaris*	51	0.461	0.178–0.214	[7]
Pectinase	*Chlorella vulgaris*	22.4	0.177	0.07	[47]

Even though alkali and acid-alkaline hydrolyze are quicker, easier, and less expensive than other methods of hydrolysis, acidic conditions can cause sugar breakdown, resulting in undesirable chemicals that inhibit fermentation [48]. Furthermore, alkaline treatment pollutes the environment and necessitates costly wastewater treatment [49]. It is environmentally friendly and can provide larger glucose yields without the production of inhibitory compounds associated with enzymatic hydrolysis [7]. As a result, algal researchers have shifted their focus to the enzymatic treatment of algal biomass for bioethanol generation. Jelynne et al. [50] used cellulase enzyme to saccharify brown seaweed (*Sargassum* sp.), but the glucose yield was only 49.84% after 72 h, and bioethanol was effectively produced by *S. cerevisiae*. Kim et al. [47] acquired the maximum saccharification yield of 79% from C. vulgaris utilizing the *Aspergillus* Pectinase enzyme, which was higher than the yield obtained by bead milling, which was only 70%.

7.6 Fermentation of Algal Sugars

Polysaccharides found in macroalgae include glucan, which is all glucose-based polysaccharides. Sulfated polysaccharides, such as agar, carrageenan, and alginate, are examples of non-glucan polysaccharides. Glucan and non-glucan hydrolysis or the subsequent fermentation of the resultant carbohydrates are necessary for the production of higher bioethanol [51]. Yeast, bacteria, and fungi have been used to ferment the monosaccharide produced during the pretreatment process, resulting in the production of bioethanol as a byproduct. *S. cerevisiae* is the most common yeast microorganism used to ferment sugar. Separate hydrolysis and fermentation (SHF) and simultaneous saccharification and fermentation (SSF) are the two most common fermentation processes. The SHF technique was used by Harun et al. [46] to manufacture bioethanol from *C. infusionum*. After alkali digestion of microalgae biomass, fermentation with ordinary yeast *S. cerevisiae* was carried out for 72 h at 30 °C and 200 rpm. The biomass pretreatment with 0.75% (w/v) sodium hydroxide at 120 °C for 30 min yielded the greatest ethanol yield of 26.1 wt.% (g ethanol/g microalgae). SSF would need less enzyme loading and produce a higher ethanol yield than SHF. SSF has a lot of advantages, like not needing separate reactors for saccharification and fermentation, a shorter fermentation time, and less risk of contamination from outside microbes [52]. Ho et al. [7] investigated the potential of the SSF method using *C. vulgaris* microalgae biomass as the substrate. Endoglucanase, β-glucosidase, and amylase were utilized as saccharifying enzymes, and *Z. mobilis* was used as the fermenting bacteria. The maximum ethanol concentration and yield were 4.27 g/L and 92.3 percent, respectively, at a biomass concentration of 20 g/L, a temperature of 30 °C, a pH of 6, and a fermentation time of 60 h.

7.7 Summary

Bioethanol production is now being explored using a wide variety of feedstocks. Algae, on the other hand, have been shown to be one of the most promising sources of ethanol. Algal bioethanol production offers significant potential for global sustainable development because of its vast distribution and ability to thrive in a variety of climates. Due to its abundance of high-quality biomolecules, macroalgae has emerged as an attractive option for biofuel production.

References

1. Dragone, G., Fernandes, B. D., Vicente, A. A., & Teixeira, J. A. (2010). Third generation biofuels from microalgae.
2. Brennan, L., & Owende, P. (2010). Biofuels from microalgae—a review of technologies for production, processing, and extractions of biofuels and co-products. *Renewable and Sustainable Energy Reviews, 14*(2), 557–577.
3. Nigam, P. S., & Singh, A. (2011). Production of liquid biofuels from renewable resources. *Progress in Energy and Combustion Science, 37*(1), 52–68.
4. Chisti, Y. (2007). Biodiesel from microalgae. *Biotechnology Advances, 25*(3), 294–306.
5. Sirajunnisa, A. R., & Surendhiran, D. (2016). Algae–A quintessential and positive resource of bioethanol production: A comprehensive review. *Renewable and Sustainable Energy Reviews, 66*, 248–267.
6. da Silva Cardoso, A., Vieira, G. E. G., & Marques, A. K. (2011). O uso de microalgas para a obtenção de biocombustíveis. *Revista Brasileira de Biociências, 9*(4), 542.
7. Ho, S. H., Huang, S. W., Chen, C. Y., Hasunuma, T., Kondo, A., & Chang, J. S. (2013). Bioethanol production using carbohydrate-rich microalgae biomass as feedstock. *Bioresource Technology, 135*, 191–198.
8. Demirbas, A., & Demirbas, M. F. (2010). *Algae energy: algae as a new source of biodiesel*. Springer Science & Business Media.

9. Ullah, K., Ahmad, M., Sharma, V. K., Lu, P., Harvey, A., Zafar, M., & Sultana, S. (2015). Assessing the potential of algal biomass opportunities for bioenergy industry: A review. *Fuel, 143*, 414–423.
10. Zhan, J., Rong, J., & Wang, Q. (2017). Mixotrophic cultivation, a preferable microalgae cultivation mode for biomass/bioenergy production, and bioremediation, advances and prospect. *International Journal of Hydrogen Energy, 42*(12), 8505–8517.
11. John, R. P., Anisha, G. S., Nampoothiri, K. M., & Pandey, A. (2011). Micro and macroalgal biomass: A renewable source for bioethanol. *Bioresource Technology, 102*(1), 186–193.
12. Chen, P., Min, M., Chen, Y., Wang, L., Li, Y., Chen, Q., Wang, C., Wan, Y., Wang, X., Cheng, Y., & Deng, S. (2010). Review of biological and engineering aspects of algae to fuels approach. *International Journal of Agricultural and Biological Engineering, 2*(4), 1–30.
13. Carlsson, A. S., Van Beilen, J. B., Möller, R., & Clayton, D. (2007). Micro-and macro-algae: utility for industrial applications. *Outputs from the EPOBIO project*, p. 82.
14. Subhadra, B., & Edwards, M. (2010). An integrated renewable energy park approach for algal biofuel production in United States. *Energy Policy, 38*(9), 4897–4902.
15. Shilton, A. N., Mara, D. D., Craggs, R., & Powell, N. (2008). Solar-powered aeration and disinfection, anaerobic co-digestion, biological CO2 scrubbing and biofuel production: The energy and carbon management opportunities of waste stabilisation ponds. *Water Science and Technology, 58*(1), 253–258.
16. Adams, J. M., Gallagher, J. A., & Donnison, I. S. (2009). Fermentation study on Saccharina latissima for bioethanol production considering variable pre-treatments. *Journal of Applied Phycology, 21*(5), 569–574.
17. Nobe, R., Sakakibara, Y., Fukuda, N., Yoshida, N., Ogawa, K., & Suiko, M. (2003). Purification and characterization of laminaran hydrolases from Trichoderma viride. *Bioscience, Biotechnology, and Biochemistry, 67*(6), 1349–1357.
18. Horn, S. J., Aasen, I. M., & Østgaard, K. (2000). Production of ethanol from mannitol by Zymobacter palmae. *Journal of Industrial Microbiology and Biotechnology, 24*(1), 51–57.
19. Hong, I. K., Jeon, H., & Lee, S. B. (2014). Comparison of red, brown and green seaweeds on enzymatic saccharification process. *Journal of Industrial and Engineering Chemistry, 20*(5), 2687–2691.
20. Fasahati, P., Woo, H. C., & Liu, J. J. (2015). Industrial-scale bioethanol production from brown algae: Effects of pretreatment processes on plant economics. *Applied Energy, 139*, 175–187.
21. Rajkumar, R., Yaakob, Z., & Takriff, M. S. (2014). Potential of micro and macro algae for biofuel production: A brief review. *BioResources, 9*(1), 1606–1633.
22. Khan, S. A., Hussain, M. Z., Prasad, S., & Banerjee, U. C. (2009). Prospects of biodiesel production from microalgae in India. *Renewable and Sustainable Energy Reviews, 13*(9), 2361–2372.
23. Mata, T. M., Martins, A. A., & Caetano, N. S. (2010). Microalgae for biodiesel production and other applications: A review. *Renewable and Sustainable Energy Reviews, 14*(1), 217–232.
24. Qiu, R., Gao, S., Lopez, P. A., & Ogden, K. L. (2017). Effects of pH on cell growth, lipid production and CO_2 addition of microalgae Chlorella sorokiniana. *Algal Research, 28*, 192–199.
25. He, L., Chen, Y., Wu, X., Chen, S., Liu, J., & Li, Q. (2020). Effect of physical factors on the growth of Chlorella vulgaris on enriched media using the methods of orthogonal analysis and response surface methodology. *Water, 12*(1), 34.
26. Bibi, R., Ahmad, Z., Imran, M., Hussain, S., Ditta, A., Mahmood, S., & Khalid, A. (2017). Algal bioethanol production technology: A trend towards sustainable development. *Renewable and Sustainable Energy Reviews, 71*, 976–985.
27. Suganya, T., Varman, M., Masjuki, H. H., & Renganathan, S. (2016). Macroalgae and microalgae as a potential source for commercial applications along with biofuels production: A biorefinery approach. *Renewable and Sustainable Energy Reviews, 55*, 909–941.
28. Ashokkumar, V., & Rengasamy, R. (2012). Mass culture of Botryococcus braunii Kutz. under open raceway pond for biofuel production. *Bioresource Technology, 104*, 394–399.

29. Bajhaiya, A. K., Mandotra, S. K., Suseela, M. R., Toppo, K., & Ranade, S. (2010). Algal biodiesel: The next generation biofuel for India. *Asian Journal of Experimental Biological Sciences, 4*, 728–739.
30. Pittman, J. K., Dean, A. P., & Osundeko, O. (2011). The potential of sustainable algal biofuel production using wastewater resources. *Bioresource Technology, 102*(1), 17–25.
31. Katarzyna, L., Sai, G., & Singh, O. A. (2015). Non-enclosure methods for non-suspended microalgae cultivation: Literature review and research needs. *Renewable and Sustainable Energy Reviews, 42*, 1418–1427.
32. Doe, U. S. (2010). National algal biofuels technology roadmap. *US Department of Energy, Office of Energy Efficiency and Renewable Energy, Biomass Program.*
33. Uduman, N., Qi, Y., Danquah, M. K., Forde, G. M., & Hoadley, A. (2010). Dewatering of microalgal cultures: A major bottleneck to algae-based fuels. *Journal of Renewable and Sustainable Energy, 2*(1), 012701.
34. Wang, Z. T., Ullrich, N., Joo, S., Waffenschmidt, S., & Goodenough, U. (2009). Algal lipid bodies: Stress induction, purification, and biochemical characterization in wild-type and starchless Chlamydomonas reinhardtii. *Eukaryotic Cell, 8*(12), 1856–1868.
35. Grima, E. M., Belarbi, E. H., Fernández, F. A., Medina, A. R., & Chisti, Y. (2003). Recovery of microalgal biomass and metabolites: Process options and economics. *Biotechnology Advances, 20*(7–8), 491–515.
36. Singh, G., & Patidar, S. K. (2018). Microalgae harvesting techniques: A review. *Journal of Environmental Management, 217*, 499–508.
37. Lee, D. H., Bae, C. Y., Han, J. I., & Park, J. K. (2013). In situ analysis of heterogeneity in the lipid content of single green microalgae in alginate hydrogel microcapsules. *Analytical Chemistry, 85*(18), 8749–8756.
38. Divakaran, R., & Pillai, V. S. (2002). Flocculation of river silt using chitosan. *Water Research, 36*(9), 2414–2418.
39. Giovannoni, S. J., DeLong, E. F., Schmidt, T. M., & Pace, N. R. (1990). Tangential flow filtration and preliminary phylogenetic analysis of marine picoplankton. *Applied and Environmental Microbiology, 56*(8), 2572–2575.
40. Bosma, R., Van Spronsen, W. A., Tramper, J., & Wijffels, R. H. (2003). Ultrasound, a new separation technique to harvest microalgae. *Journal of Applied Phycology, 15*(2), 143–153.
41. Gröschl, M. (1998). Ultrasonic separation of suspended particles-Part I: Fundamentals. *Acta Acustica United with Acustica, 84*(3), 432–447.
42. Schenk, P. M., Thomas-Hall, S. R., Stephens, E., Marx, U. C., Mussgnug, J. H., Posten, C., Kruse, O., & Hankamer, B. (2008). Second generation biofuels: High-efficiency microalgae for biodiesel production. *Bioenergy Research, 1*(1), 20–43.
43. Munoz, R., & Guieysse, B. (2006). Algal–bacterial processes for the treatment of hazardous contaminants: A review. *Water Research, 40*(15), 2799–2815.
44. Judge, D., & Earnshaw, D. (2003). *The European Parliament.* Palgrave: Basingstoke, UK.
45. Miranda, J. R., Passarinho, P. C., & Gouveia, L. (2012). Pre-treatment optimization of Scenedesmus obliquus microalga for bioethanol production. *Bioresource Technology, 104*, 342–348.
46. Harun, R., Jason, W. S. Y., Cherrington, T., & Danquah, M. K. (2011). Exploring alkaline pre-treatment of microalgal biomass for bioethanol production. *Applied Energy, 88*(10), 3464–3467.
47. Kim, K. H., Choi, I. S., Kim, H. M., Wi, S. G., & Bae, H. J. (2014). Bioethanol production from the nutrient stress-induced microalga Chlorella vulgaris by enzymatic hydrolysis and immobilized yeast fermentation. *Bioresource Technology, 153*, 47–54.
48. Harun, R., Danquah, M. K., & Forde, G. M. (2010). Microalgal biomass as a fermentation feedstock for bioethanol production. *Journal of Chemical Technology & Biotechnology, 85*(2), 199–203.
49. Surendhiran, D., Sirajunnisa, A. R., & Vijay, M. (2015) An alternative method for production of microalgal biodiesel using novel Bacillus lipase. *3 Biotech, 5*(5), 715–725.

50. Tamayo, P., & Del Rosario, E. J. (2014). Chemical analysis and utilization of Sargassum sp. as substrate for ethanol production. *Iranian (Iranica) Journal of Energy & Environment*, *5*(2).
51. Kumar, S., Gupta, R., Kumar, G., Sahoo, D., & Kuhad, R. C. (2013). Bioethanol production from Gracilaria verrucosa, a red alga, in a biorefinery approach. *Bioresource Technology, 135*, 150–156.
52. Lin, Y., & Tanaka, S. (2006). Ethanol fermentation from biomass resources: Current state and prospects. *Applied Microbiology and Biotechnology, 69*(6), 627–642.

Chapter 8
Integrated Biorefinery and Bioethanol Production

Abstract The need for alternative renewable resources has grown in response to rising energy demand and depleted petroleum sources. Integrated biorefinery is an ideal approach for this where production of bioethanol can be done along with different products. Though no doubt, there are many shortcomings in the way like higher production costs but development and advancements in different technical methods of saccharification and recovery can overcome such hurdles. This chapter focuses on the concept of biorefinery, different types of biorefineries, and production of bioethanol.

Keywords Biorefinery · Bioethanol · Pretreatment · Saccharification

8.1 Introduction

Fossil fuels are indeed major sources of both chemical and energy, approximately 75% are utilized for energy or heat generation, 20% for fuel, and only a small percent for chemical and material manufacturing [1]. The rate at which fossil resources regenerate naturally with the aid of the carbon cycle is substantially slower than the rate at which they are currently used. The largest deposits of fossil fuels are held by a limited number of nations, increasing the unsustainable nature of their production. Furthermore, rising greenhouse gas emissions are caused by human activities such as fossil fuel burning and land-use change, resulting in a worsening of the global warming situation [2]. Governments in most developed nations encourage the use of renewable energy and resources with the following key objectives: (i) ensuring energy access, (ii) mitigating climate change, (iii) developing/maintaining agricultural operations, and (iv) ensuring food safety. Replacement of fossil resources with renewable alternatives, which are comparatively more evenly distributed and create lesser environmental as well as social apprehensions, could successfully improve the current scenario of global warming and all fossil-based problems [3].

N. K. Aggarwal et al., *Bioethanol Production*, Green Chemistry and Sustainable Technology, https://doi.org/10.1007/978-3-031-05091-6_8

Lignocellulosic biomass has the potential to reduce global warming by serving as a renewable and alternative energy source. The use of biomass wastes for the synthesis of high-value chemicals can help to mitigate the negative impacts of burning on soil quality and the environment. On the other side, it has the potential to boost rural economies while also promoting energy security. However, due to the very complicated arrangement and recalcitrance nature of lignocellulosic feedstock, cost-effective conversion into renewable sources or biofuels and value-added products is extremely difficult. Valorization of lignocellulosic biomass leftovers produced by agricultural, forestry, and industrial operations for biofuels and other value-added chemicals synthesis in a biorefinery is gaining attention and therefore is becoming important sector of research. By turning biomass into valuable compounds and bioenergy in a combined fashion, the economic value of the biomass can be increased while also reducing waste streams [4]. Biorefining lignocellulosic waste into second-generation (2G) bioethanol and other value-added products are not only good for the environment, but also promotes long-term development. Under the biorefinery regime, three key components of lignocellulosic feedstock, cellulose, hemicellulose, and lignin, are promising players in the growth of the global bioeconomy [5]. More than 200 value-added biochemicals can be produced from lignocellulosic biorefinery [6].

Biorefinery is a wide term that encompasses diverse categories, like, chemical, transport, energy, and agriculture, and therefore it is not easy to define in a single way [7]. For example, according to the International Energy Agency (IEA), biorefining is the sustainable conversion of biomass into broad range of bio-based products and biofuels, as an efficient scheme to employ available feedstock for the synergistic production of energy. According to another definition, biorefinery is defined as "a facility for the synergetic conversion of biomass into multiple commercial bio-based products (food and feed ingredients, chemicals, materials, minerals, CO_2) and bioenergy (fuels, power, heat)" [8]. Biorefineries are, more broadly, linked to the concept of a circular economy, which has been recognized as a key feature for enhancing the efficiency and sustainability of natural resource production and utilization in the near future, as a response to the global increase in product and energy demand (Fig. 8.1).

Fig. 8.1 Different components of integrated biorefinery

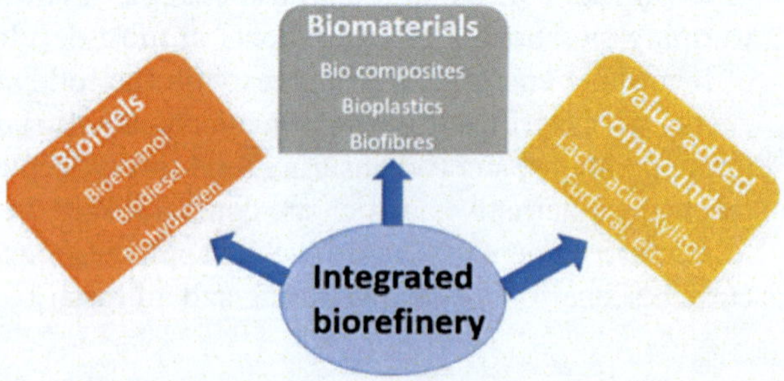

8.2 Concept of Biorefinery

Pretreatment and processing of substrate, separation of organic substances, and successive conversion and product recovery phases are typically included in the biorefinery process. Bottom-up and top-down approaches to biorefinery idea implementation are the two main techniques. The bottom-up approach is defined by the integration of existing biomass processing facilities (producing only one or a few products) into a biorefinery toward the goal of expanding the range of products and/or increasing the amount of usable biomass fractions by connecting to additional techniques. The wheat and maize starch biorefinery, which began as a simple starch mill, is a case of bottom-up biorefinery. It gradually increased the number of products it offered, such as starch derivatives and alterations, chemicals, and fermentation products. Bottom-up approaches are also used by a corn starch biorefinery in the United States and wood lignocellulosic biorefineries in Austria and Norway [9]. The new top-down strategy is a highly combined system for the usage of diverse biomass fractions and the production of various marketable products (zero-waste generation). The goal is to utilize all of the biomass available (e.g., wood lignocellulosic biomass, and straw obtained from cereals). Austrian Green Biorefinery is an example of a top-down method. Grass silage is used as a biomass for the manufacture of bio-based products such as lactic acid, proteinaceous compounds, fibers, and biogas. Green grass juice and silage juice were also used as cultural medium elements for *Wautersia eutropha* growth and PHA synthesis [10]. Top-down biorefineries are still in the early phases of development, with demonstration plants mostly in the United States, Europe, and other industrialized countries [11]. Figure 8.2 shows both bottom-up and top-down biorefinery principles.

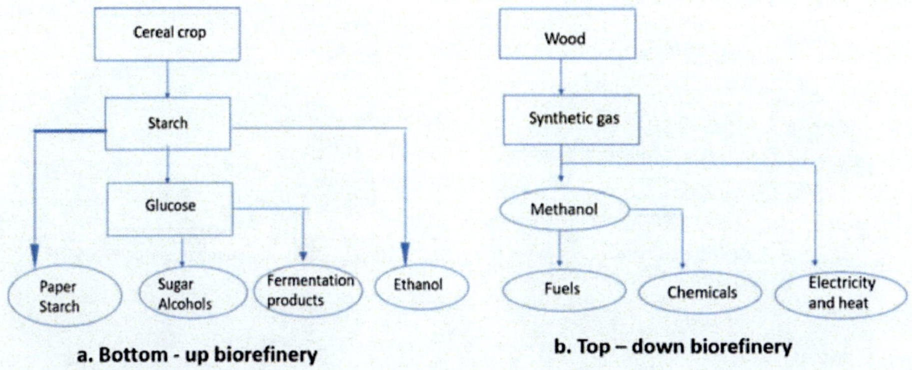

Fig. 8.2 Concept of bottom-up and top-down biorefineries

8.3 Bioethanol

Bioethanol is primarily created by yeast fermentation from various feedstocks as an alternative to fossil fuels [12]. Bioethanol is primarily used in the transportation industry as a component of gasoline blends or as an octane booster (ethyl tertiary butyl ether (ETBE), which contains 45 percent bioethanol and 55 percent isobutylene). Many nations utilize ETBE rather than methyl tertiary butyl ether (MTBE), which increases octane number but is banned in the United States and also in Canada due to carcinogenic discharges. Bioethanol is blended with gasoline at percentages of 5, 10, and 85 percent (fuel names E5-E85). Only flexible-fuel vehicles (FFV) can use a total of 85 percent bioethanol by volume, while combinations of 5 and 10% by volume can be used without any engine changes. However, there are a number of issues with using bioethanol, including corrosive effects on fuel injectors and electric fuel pumps (bioethanol is hygroscopic), engine startup issues in cold conditions (pure ethanol is difficult to vaporize), and tribological effects on lubricant characteristics and engine performance. Bioethanol in the lubricant affects the characteristics and performance of engine oil significantly. It is miscible in water but not in oil. As a result, bioethanol has a significant capability for forming emulsions (mixtures of bioethanol, water, and oil), which can cause serious engine failures. There are various methods for improving engine performance (e.g., coatings, reduction in mass of engine parts, laser texturing, and lubricant composition) and extending their period by reducing friction and wear. One option for resolving the aforementioned concerns is to utilize synthetic oil [13]. Global bioethanol production in 2016 was 100.2 billion liters, according to data [14]. Annual bioethanol output is steadily expanding, with global bioethanol production and consumption expected to reach roughly 134.5 billion liters by 2024 (Fig. 8.3) [15].

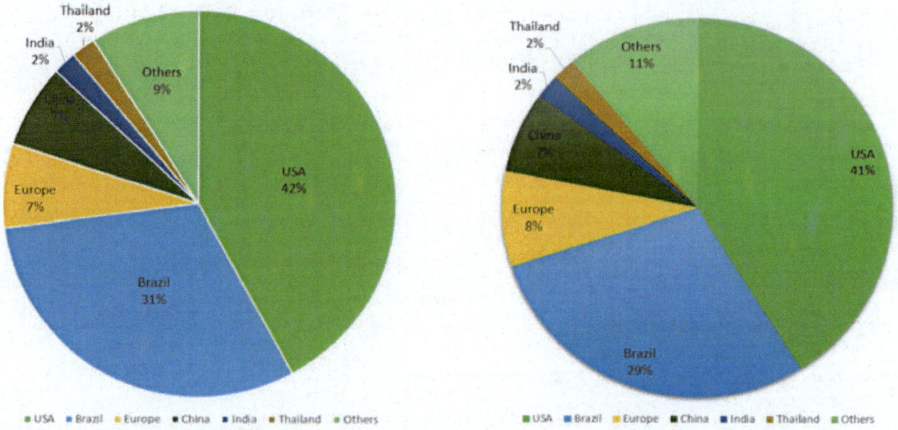

Fig. 8.3 Predictions of global bioethanol production as well as consumption by 2024

8.4 Lignocellulosic Biorefinery for Production of Bioethanol

Lignocellulosic feedstock can be categorized into different groups like energy crops (perennial grass), agriculture residual waste (bagasse, straws, etc.), forest material (softwood, hardwood, bark residues, etc.), and organic fraction of municipal solid waste [16]. cellulose (35–50%), hemicelluloses (20–35%), lignin (5–30%), and other extracted compounds (1–10%) are the main elements of lignocellulosic biomass [17].

8.4.1 Pretreatment of LCB Feedstock

Pretreatment is required for the hydrolysis of refractory lignocellulose's closely interwoven matrix. To make the feedstock more accessible to the subsequent digestion processes to liberate the fermentable sugars, lignin must be removed, hemicellulose must be partially or completely hydrolyzed, the crystalline section of cellulose must be reduced, and the degree of polymerization must be reduced. There are a variety of pretreatment processes like physical, chemical, physicochemical, and most important biological that promote lignocellulose digestibility in various ways [18]. The requirement for selecting a pretreatment process in a 2G biorefinery is the production of easily hydrolysable cellulose-rich biomass. The assessment of cellulose hydrolystability by cellulases can be used to choose a pretreatment technique for enzymatic hydrolysis. Using 10 FPU/g of enzyme, a study predicted that the pretreatment approach should allow more than 90 percent of cellulose to be hydrolyzed from pretreated biomass in less than three days [19]. Furthermore, many factors such as the chemical composition of the feedstock, the loss of feedstock, the production of fermentation or enzyme inhibitors, the consumption of water, the degradation of sugars, the ease with which lignin and other valuables can be recovered, economic feasibility, environmental safety, and so on must all be considered when selecting a pretreatment process. Biomass is normally separated into its structural parts, cellulose, hemicellulose, and lignin, in a lignocellulosic biorefinery. This step is critical for each polymer's further valorization. In an integrated biorefinery, the fractionation process should be able to distinguish each component as well as assist fraction recovery procedures with minimal purifying stages. Dilute acid, hydrothermal, alkali, steam, and organosolv are some of the successful fractionation techniques. Pre-extraction and separation of hemicelluloses and lignin, followed by the manufacture of value-added compounds like ethanol, polyesters, other chemical compounds, and biopolymers, offer a potential possibility in a combined 2G ethanol biorefinery [20]. Table 8.1 shows a quantitative comparative account of some of the important pretreatment processes.

Table 8.1 Different pretreatment methods

Pretreatment method	Efficiency
Mechanical pretreatment	Helps in liberation of sugars after enzymatic saccharification of rice straw:89% [26]
Alkaline Pretreatment	Helps in delignification of rice straw: 64.51% [27], Liberation of sugars after enzymatic digestion of sugarcane: 97.6% [28]
Ionic Liquids	Sugar liberation after hydrolysis of triticale straw, i.e., 81% [29], Delignification of corn straw at the rate of 81.73% [30]
Steam explosion	Delignification of barley straw 85% [31], glucan conversion of corn stover, i.e., 87% [32]
Ammonia fiber explosion (AFEX)	Delignification of corn stover rate 24% [33], Xylan conversion of corn stover, i.e., 91% [32]
Liquid hot water	Recovery of sugar up to 95% with higher enzyme efficiency of loblolly pine [34], Delignification of sugarcane bagasse: 77% [35]

8.4.2 Enzymatic Hydrolysis

Saccharification of both cellulosic and hemicellulosic fraction of pretreated feedstock is the most important step which could affect the economics of the conversion process. Enzymatic saccharification is regarded as most efficient, viable, specific, viable, and also eco-friendly method over nonenzymatic method. Additionally, digestion of cellulosic biomass with the aid of cellulases enzyme does not produce inhibitors. However, gradual advances in pretreatment have led to many folds reduction in enzyme loading for the purpose of degradation, the effectual enzymatic conversion of cellulose and hemicellulose is still a challenging task in using biomass for production of biofuel and other value-added compounds. Cellulases are a combination of at least three enzymes generated primarily from fungi: I Endoglucanase or beta–1,4–D–glucan–4– glucanohydrolases (EC 3.2.1.4) randomly cleave the interior glycosidic connections in cellulose, resulting in oligosaccharides of varying chain lengths and the opening of new chain ends. (ii) Exoglucanases, such as –1,4–D–glucan glucanohydrolases (EC 3.2.1.74) and –1,4–D–glucan cellobiohydrolases (EC 3.2.1.91), act in a processive fashion on the reducing and nonreducing ends of cellulose polymeric chains, liberating glucose (glucanohydrolases) or cellobiose (cellobiohydrolases) (cellobiohydrolase). (iii) –D–Glucosidases (EC 3.2.1.21), which operate on cellobiose and cellodextrins to liberate D-glucose units [21]. For complete hydrolysis of cellulose to glucose, a combination of enzymes (endoglucanase, exoglucanase, and –glucosidase) is required, which functions synergistically for both native and modified cellulose hydrolysis [22]. Main-chain hydrolyzing enzymes (mannanase, xylanase, xylosidase, mannosidase) and side-chain-cleaving enzymes (esterases, –L–arabinofuranosidase) are two types of hemicellulases. Apart from cellulolytic enzymes, ligninolytic enzymes are also important for removing lignin and so facilitating biomass enzymatic saccharification. The major enzymes in lignin breakdown are manganese peroxidase (MnP), and lignin peroxidase (LiP) [23]. Laccases use molecular oxygen as an end electron acceptor to oxidize aromatic amines and phenolic chemicals. At the non-phenolic and phenolic aryl-ether sites, the MnP and LiP can oxidize lignin, respectively [24]. Recent research has sparked a lot of interest in using laccase-cellulase cocktails to pretreat and saccharify lignocellulosic biomass at the same time [25].

8.4.3 Fermentation

Separate hydrolysis and fermentation (SHF), Simultaneous Saccharification and Fermentation (SSF), Co-fermenation (CF), Simultaneous Saccharification and Co-fermentation (SSCF), and Consolidated Bioprocessing are some of the fermentation processes available for producing 2G ethanol from lignocellulosic biomass (CBP). The SHF technique creates the ideal environment for enzymatic saccharification and ethanologen fermentation. The effect of glucose and cellobiose on cellulase activity is, however, a key downside of the SHF process. The SSF process is more cost-effective and produces more ethanol because both enzymatic hydrolysis and fermentation take place at the same time, avoiding the difficulties of cellulase feedback inhibition. However, because the optimum temperature of enzyme reactions is substantially higher than the temperature at which ethanologens ferment, enzymatic digestion becomes a rate-limiting step in the SSF process [36]. At high solid loading, the SSF process delivers significant cost savings as well as a high concentration of ethanol, which is desired for distillation [37]. High solid loading, on the other hand, causes the medium to become viscous, resulting in increased levels of inhibitory chemicals and significant mixing issues [38]. To solve these disadvantages, prehydrolysis, simultaneous saccharification, and co-fermentation were planned within the integrated biorefinery [39]. The SSCF is similar to the SSF in that it involves the fermentation of both hexoses and pentoses, resulting in a higher ethanol production. In comparison to the SHF and CF processes, the SSCF involves simultaneous fermentation of the liberated glucose, which reduces the inhibition imposed by the end product. Furthermore, it raises the xylose-to-glucose ratio, allowing fermenting microorganisms to utilize xylose [40]. The CBP is a promising technique that uses engineered microbes to create cellulases and ethanol in a single step, lowering the overall cost of the bioconversion process. *Clostridium thermocellum* and *Clostridium cellulolyticum* are being evaluated for high-efficiency CBP employing gene transfer research and organism development procedures [41]. Consolidated Bio-Saccharification (CBS), a new CBP-derived technique for lignocellulose bioconversion, has recently been proposed, in which enzyme production is linked with hydrolysis phases but isolated from the fermentation process. As a result, cellulolytic capability can be increased, and fermentation will not be constrained by hydrolysis conditions [42].

8.5 Summary

The significant investments made recently in research and development for enhanced production of bioethanol from lignocellulosic feedstock via biochemical conversion have resulted in this biofuel being a vital factor in fulfilling future bioethanol demand. To reach full commercialization of cellulosic ethanol, however, higher production yields and lower costs are still required. To attain this goal, it has been suggested that the use of novel and enhanced approaches in the various stages of the conversion methods is required in order to gain more effective exploitation of lignocellulosic biomass sources while also increasing the overall process's sustainability. In this regard, the researchers have been working diligently to offer technological advancements at every stage of the lignocellulosic feedstock to bioethanol value chain, from the development of new and improved catalysts and microbes in the final conversion process steps to the use of alternative and promising biomass sources. Furthermore, within a biorefinery approach, the introduction of advanced technologies to obtain new value-added compounds resulting from a more efficient and sustainable use of biomass has been recognized as a very promising way to increase the production of biofuels such as bioethanol and other bio-based products. Finally, industrial developers must test and verify that these new methodologies are effective, which is unquestionably a difficult task.

References

1. Bhaskar, T., Bhavya, B., Singh, R., Naik, D. V., Kumar, A., & Goyal, H. B. (2011). Thermochemical conversion of biomass to biofuels. In *Biofuels* (pp. 51–77). Academic Press.
2. Forster, P., Ramaswamy, V., Artaxo, P., Berntsen, T., Betts, R., Fahey, D.W., Haywood, J., Lean, J., Lowe, D.C., Myhre, G., & Nganga, J. (2007). Changes in atmospheric constituents and in radiative forcing. Chapter 2. In *Climate change 2007. The physical science basis.*
3. Pandey, A., Larroche, C., & Ricke, S. C. (Eds.). (2011). *Biofuels: alternative feedstocks and conversion processes.* Academic Press.
4. Thomsen, M. H. (2005). Complex media from processing of agricultural crops for microbial fermentation. *Applied Microbiology and Biotechnology, 68*(5), 598–606.
5. Chandel, A. K., Garlapati, V. K., Singh, A. K., Antunes, F. A. F., & da Silva, S. S. (2018). The path forward for lignocellulose biorefineries: Bottlenecks, solutions, and perspective on commercialization. *Bioresource Technology, 264*, 370–381.
6. Isikgor, F. H., & Becer, C. R. (2015). Lignocellulosic biomass: A sustainable platform for the production of bio-based chemicals and polymers. *Polymer Chemistry, 6*(25), 4497–4559.
7. Lindorfer, J., Lettner, M., Fazeni, K., Rosenfeld, D., Annevelink, B., & Mandl, M. (2019). *Technical, economic and environmental assessment of biorefinery concepts*; IEA bioenergy: Paris, France.
8. Concepts, E. B. (2017). European Technology and Innovation Platform Bioenergy: Gülzow, Germany.
9. Roadmap, B. (2012). *German Federal Government action plans for the material and energetic utilisation of renewable raw materials.*
10. Kromus, S., Wachter, B., Koschuh, W., Mandl, M., Krotscheck, C., & Narodoslawsky, M. (2004). The green biorefinery Austria-development of an integrated system for green biomass utilization. *Chemical and Biochemical Engineering Quarterly, 18*(1), 8–12.

11. Koller, M., Bona, R., Hermann, C., Horvat, P., Martinz, J., Neto, J., Pereira, L., Varila, P., & Braunegg, G. (2005). Biotechnological production of poly (3-hydroxybutyrate) with Wautersia eutropha by application of green grass juice and silage juice as additional complex substrates. *Biocatalysis and Biotransformation, 23*(5), 329–337.
12. Yüksel, F., & Yüksel, B. (2004). The use of ethanol–gasoline blend as a fuel in an SI engine. *Renewable Energy, 29*(7), 1181–1191.
13. Khuong, L. S., Masjuki, H. H., Zulkifli, N. W. M., Mohamad, E. N., Kalam, M. A., Alabdulkarem, A., Arslan, A., Mosarof, M. H., Syahir, A. Z., & Jamshaid, M. (2017). Effect of gasoline–bioethanol blends on the properties and lubrication characteristics of commercial engine oil. *RSC Advances, 7*(25), 15005–15019.
14. Kummamuru, B. (2016). WBA global bioenergy statistics 2017. *World Bioenergy Association.*
15. Outlook, O.F.A. (2010). © OECD/FAO 2010.
16. Zabed, H., Sahu, J. N., Suely, A., Boyce, A. N., & Faruq, G. (2017). Bioethanol production from renewable sources: Current perspectives and technological progress. *Renewable and Sustainable Energy Reviews, 71*, 475–501.
17. Menon, V., & Rao, M. (2012). Trends in bioconversion of lignocellulose: biofuels, platform chemicals & biorefinery concept. *Progress in energy and combustion science, 38*(4), 522–550.
18. Alvarado-Morales, M., Terra, J., Gernaey, K. V., Woodley, J. M., & Gani, R. (2009). Biorefining: Computer aided tools for sustainable design and analysis of bioethanol production. *Chemical Engineering Research and Design, 87*(9), 1171–1183.
19. Yang, B., & Wyman, C. E. (2008). Pretreatment: The key to unlocking low-cost cellulosic ethanol. *Biofuels, Bioproducts and Biorefining: Innovation for a Sustainable Economy, 2*(1), 26–40.
20. Lin, C., & Luque, R. (Eds.). (2014). *Renewable resources for biorefineries* (No. 27). Royal Society of Chemistry.
21. Shah, A., Patel, H., & Narra, M. (2017). Bioproduction of fungal cellulases and hemicellulases through solid state fermentation. *Fungal Metabolites, 349*, 393.
22. Saini, J. K., Saini, R., & Tewari, L. (2015). Lignocellulosic agriculture wastes as biomass feedstocks for second-generation bioethanol production: concepts and recent developments. *3 Biotech, 5*(4), 337–353.
23. Chukwuma, O. B., Rafatullah, M., Tajarudin, H. A., & Ismail, N. (2020). Lignocellulolytic enzymes in biotechnological and industrial processes: A review. *Sustainability, 12*(18), 7282.
24. Lopes, A. D. M., Ferreira Filho, E. X., & Moreira, L. R. S. (2018). An update on enzymatic cocktails for lignocellulose breakdown. *Journal of Applied Microbiology, 125*(3), 632–645.
25. Masran, R., Bahrin, E. K., Ibrahim, M. F., Phang, L. Y., & Abd-Aziz, S. (2020). Simultaneous pretreatment and saccharification of oil palm empty fruit bunch using laccase-cellulase cocktail. *Biocatalysis and Agricultural Biotechnology, 29*, 101824.
26. Hideno, A., Inoue, H., Tsukahara, K., Fujimoto, S., Minowa, T., Inoue, S., Endo, T., & Sawayama, S. (2009). Wet disk milling pretreatment without sulfuric acid for enzymatic hydrolysis of rice straw. *Bioresource Technology, 100*(10), 2706–2711.
27. Rezania, S., Oryani, B., Cho, J., Talaiekhozani, A., Sabbagh, F., Hashemi, B., Rupani, P. F., & Mohammadi, A. A. (2020). Different pretreatment technologies of lignocellulosic biomass for bioethanol production: An overview. *Energy, 199*, 117457.
28. Nosratpour, M. J., Karimi, K., & Sadeghi, M. (2018). Improvement of ethanol and biogas production from sugarcane bagasse using sodium alkaline pretreatments. *Journal of environmental management, 226*, 329–339.
29. Smuga-Kogut, M., Walendzik, B., Szymanowska-Powalowska, D., Kobus-Cisowska, J., Wojdalski, J., Wieczorek, M., & Cielecka-Piontek, J. (2019). Comparison of bioethanol preparation from triticale straw using the ionic liquid and sulfate methods. *Energies, 12*(6), 1155.
30. Liu, X., Liu, Y., Jiang, Z., Liu, H., Yang, S., & Yan, Q. (2018). Biochemical characterization of a novel xylanase from Paenibacillus barengoltzii and its application in xylooligosaccharides production from corncobs. *Food Chemistry, 264*, 310–318.

31. Oliva, J. M., Negro, M. J., Manzanares, P., Ballesteros, I., Chamorro, M. Á., Sáez, F., Ballesteros, M., & Moreno, A. D. (2017). A sequential steam explosion and reactive extrusion pretreatment for lignocellulosic biomass conversion within a fermentation-based biorefinery perspective. *Fermentation, 3*(2), 15.
32. da Costa Sousa, L., Chundawat, S. P., Balan, V., & Dale, B. E. (2009). 'Cradle-to-grave'assessment of existing lignocellulose pretreatment technologies. *Current Opinion in Biotechnology, 20*(3), 339–347.
33. Zhao, C., Ding, W., Chen, F., Cheng, C., & Shao, Q. (2014). Effects of compositional changes of AFEX-treated and H-AFEX-treated corn stover on enzymatic digestibility. *Bioresource Technology, 155*, 34–40.
34. Biswas, R., Uellendahl, H., & Ahring, B. K. (2015). Wet explosion: A universal and efficient pretreatment process for lignocellulosic biorefineries. *BioEnergy Research, 8*(3), 1101–1116.
35. Gurgel, L. V. A., Pimenta, M. T. B., & da Silva Curvelo, A. A. (2016). Ethanol–water organosolv delignification of liquid hot water (LHW) pretreated sugarcane bagasse enhanced by high–pressure carbon dioxide (HP–CO2). *Industrial crops and products, 94*, 942–950.
36. Patel, A., Patel, H., Divecha, J., & Shah, A. R. (2019). Enhanced production of ethanol from enzymatic hydrolysate of microwave-treated wheat straw by statistical optimization and mass balance analysis of bioconversion process. *Biofuels.*
37. Hoyer, K., Galbe, M., & Zacchi, G. (2013). The effect of prehydrolysis and improved mixing on high-solids batch simultaneous saccharification and fermentation of spruce to ethanol. *Process Biochemistry, 48*(2), 289–293.
38. López-Linares, J. C., Romero, I., Cara, C., Castro, E., & Mussatto, S. I. (2018). Xylitol production by Debaryomyces hansenii and Candida guilliermondii from rapeseed straw hemicellulosic hydrolysate. *Bioresource Technology, 247*, 736–743.
39. Chen, H., & Fu, X. (2016). Industrial technologies for bioethanol production from lignocellulosic biomass. *Renewable and Sustainable Energy Reviews, 57*, 468–478.
40. Toor, M., Kumar, S. S., Malyan, S. K., Bishnoi, N. R., Mathimani, T., Rajendran, K., & Pugazhendhi, A. (2020). An overview on bioethanol production from lignocellulosic feedstocks. *Chemosphere, 242*, 125080.
41. Lynd, L. R., Van Zyl, W. H., McBride, J. E., & Laser, M. (2005). Consolidated bioprocessing of cellulosic biomass: An update. *Current Opinion in Biotechnology, 16*(5), 577–583.
42. Liu, Y. J., Li, B., Feng, Y., & Cui, Q. (2020). Consolidated bio-saccharification: Leading lignocellulose bioconversion into the real world. *Biotechnology Advances, 40*, 107535.

Chapter 9
Bioethanol: A Fermentation Feedstock for Synthesis of Bulk Chemicals

Abstract Bioethanol is being generated in greater quantities from biomass fermentation, mostly to combat the continued depletion of natural fuels and the resulting rise in oil costs. Bioethanol is now used mostly as a fuel or fuel substitute in automobiles, although it has the potential to be a flexible substrate in the chemical sector. Currently, the manufacturing of carbon-containing chemicals and fuels is reliant on fossil resources, with non-renewable carbon resources accounting for more than 95% of these chemicals. This chapter emphasizes the production of different value-added chemicals from bioethanol.

Keywords Bioethanol · Value-added chemicals · Lignocellulose · Fermentation · Economy

9.1 Introduction

Growing demand for CO_2—neutral petroleum products, as well as a desire to lessen reliance on fossil resources, have prompted a significant growth in the amount of bioethanol generated by biomass fermentation. A fascinating debate is whether bioethanol is best used as being fuel or feedstock for exclusive chemical goods. Fossil fuels and carbon-containing compounds account for the great bulk of all fuels and chemicals. Most fossil resources are expected to be depleted within the next century, according to studies [1]. Furthermore, the combustion of petroleum-based fuels results in an increase in greenhouse gases (GHGs) which may contribute to the climate change [2, 3]. As a result, society should accept transition from a fossil-fuel-based economy to the fuel-saving renewable resources one. Biomass plays a crucial role in this. Obviously, the capital of both fossil fuels and renewable materials is subject to a variety of conditions, therefore the scale is simply a rough estimate. All of the chemicals required by society can be made from these raw sources. The required conversions are usually catalytic reactions, and the pricing of the required chemicals and fuels is determined by the raw material cost and efficiency of the related processes. It is worth noting that with the right technology, biomass can be converted into virtually all of the high-value commodity chemicals and fuels that are currently available from fossil fuels. There may even be some benefits to using

N. K. Aggarwal et al., *Bioethanol Production*, Green Chemistry and Sustainable Technology, https://doi.org/10.1007/978-3-031-05091-6_9

biomass as a feedstock rather than fossil fuels, such as the manufacture of some oxygenated compounds because introducing oxygen functions into hydrocarbons can be challenging, but many biomass-derived products already contain some oxygen [4]. However, gasoline-like non-oxygenated products will see a rise in production costs as a result of this. Some compounds can be made without difficulty in fewer steps from biomass, while others are more difficult to get by [5]. As a result, not even all renewable resources are equally useful as initial compounds for all probable products. This is especially crucial to note in this context because the number does not reflect the quantity of each chemical generated. For example, it is now impossible to meet the global ethene demand with biomass because its yield is about thrice to that of ethanol and that would be the demanding fuel for ethene generation in the sustainable economy.

9.2 Bioethanol as Fermentation Feedstock

Rass-Hansen et al. have examined bioethanol as a substrate for the industrial applications, stating that bioethanol may be used to produce ethylene, acetaldehyde, 1,3-butadiene, and vinegar among other things [6]. Though, only chemical variations to the required product were described by these researchers. We look at the possibility of ethanol as being substrate for fermentation operations as an alternative transformation technique in this paper.

It has previously been investigated the usage of bioethanol as being fermentable substrate meant for the manufacture of other compounds. Interestingly, the ethanol used in these procedures was frequently derived from petrochemicals [7]. Mineral oil costs were so poor at the moment, in the 1960s and 1970s, that petroleum-based sources outcompeted agronomic sources to produce foodstuffs in terms of commercial competition. As a result, this appeared to be an appealing method of increasing food output. About 1% of overall petroleum output (2.5 billion tons each year) was enough to produce around 16 million tons of yeast protein and to meet the protein demand of approximately 2 billion people [7]. However, as the price of petrochemical resources rose after the 1970s, the strategy of converting these materials into various foods was discarded. Several firms, including Amoco Foods in the United States, Nestlé Alimentana in Switzerland, and Esso Research and Engineering in the United States, have sought to develop food items using microbial biomass produced by ethanol fermentation.

Torula yeast, made by Amoco, is an example of bacterial protein products created from bioethanol [8]. Some other food additives like organic acids and many amino acids were also investigated. The usage of bioethanol as a carbon-rich resource for the synthesis of L-glutamic acid has been investigated [9]. Accumulation of L-glutamic acid seems to be widespread among microorganisms such as *Brevibacterium* sp., *Corynebacterium* sp., and *Pseudomonas*, *Alcaligenes,* and *Bacillus* strains. The greatest concentration of L-glutamic acid produced was 53.1 g/l, yielding 0.6 g/g ethanol. The prospect of bioethanol as a carbon source for the manufacture of L-lysine was examined by the Japanese firm Kyowa Hakko Kogyo. L-lysine concentrations of 2.1–8.4 g/l were produced using *C. glutamicum, Brevibacterium ammoniagenes, Arthrobacter paraffineus, Nocardia strains, and Bacillus megaterium,* with a fairly low maximum production of 0.02 g/g ethanol [10]. One more patent from Mitsubishi Petrochemical Co Ltd claims that *Acinetobacter calcoaceticus* strain YK-1011 was able to produce a tenfold greater yield of 0.2 g L-lysine per 1-g substrate during bioethanol fermentation [11].

Yarrowia lipolytica strains were used to study the generation of citric acid, isocitric acid, and 2-oxoglutaric acid from ethanol at the G. K. Skryabin Institute in Moscow [12]. The generation of acetic acid (vinegar) is presently operating economical fermentation method dependent on ethanol, with a predictable production of approximately 500 million liters vinegar (10% acetic acid) in the European Union each year. Aeration is crucial for the generation of acetic acid. The acetic acid production is almost stoichiometrically converted [13].

9.3 Two-Stage Fermentation

These cases demonstrate that ethanol can be used as a substrate for microbial chemical synthesis, implying that lignocellulosic hydrolysates-based ethanol could also be used. This technique results in a two-stage manufacturing process: first, lignocellulose would be transformed into ethanol, then ethanol would be turned into the required product (Fig. 9.1). Vinegar manufacture is a prime example of a two-stage process, with bakers' yeast initially converting sugar to alcohol, followed by a second transformation process in which the ethanol is transformed to acetic acid by, for example, *Acetobacter* species [13].

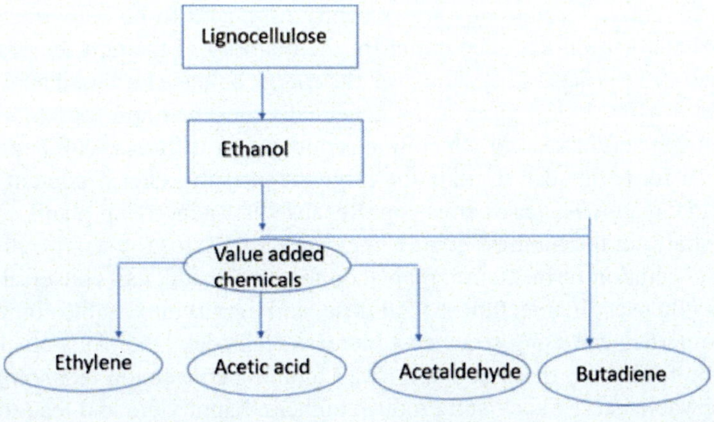

Fig. 9.1 Production of different compounds from ethanol

9.4 How Does Ethanol Have Potential to Act as a Fermentation Feedstock?

The cost of feedstocks is a major consideration when comparing the utility of fossil resources vs renewable resources as potential resources of chemical compounds and fuels. As a result, it is worth noting that the price of biomass has dropped more than tenfold in comparison to oil over the last 60 years. The pricing of items made using sustainable resources is likely more responsive to the technical advancements in manufacturing process because the processes used turned out to be improved over so many years. Finding more effective methods for converting biomass to chemicals or fuels would be a significant advance [14]. If the corn to oil cost ratio remains in falling situation and technology advances, the relative cost of biomass-derived fuels and chemicals will fall and it makes the conversion to biomass more practical. Furthermore, because biomass can be cultivated in almost any part of the world, using biomass as a sustainable resource for the manufacture of fuels and commodity chemicals could increase supply security. In contrast, oil resources are limited due to their few locations and quite unstable locations on earth.

Alcohol is a chemical compound that can simply be produced from feedstock through fermentation process. Global bioethanol output is steadily expanding, with estimates of more than 46 billion liters per year in 2005. Ethanol is utilized as a fuel additive in many nations. Brazil and the United States, in particular, have made major investments in the use of ethanol-gasoline blends as a car fuel (with internal combustion engines). Bioethanol currently accounts for about 2% of total transportation fuels in the United States [15]. By 2025, the US Department of Energy wants to replace 30% of transportation fuels with bioethanol and biodiesel [15]. Many other countries are making or proposing to make ethanol from biomass for use as a vehicle fuel. By December 2010, the European Union has set a target of 5.75% biomass-derived transportation fuels [15]. The renewable energy content for producing bioethanol from maize with current technology was recently estimated to be between 5 and 26%, with coal and natural gas accounting for the majority of the non-renewable energy consumed in the process [16]. However, the energy balance for bioethanol generation is now in dispute. In any case, using genetically modified non-food biomass that is rich in starch and/or second-generation fermentation facilities with better lignocellulose transformation can result in the surge in renewable energy content and a large decline in Greenhouse gases emission. Because first-generation plants consume the carbohydrates in the biomass source, they are only useful for starch-rich materials. Second-generation fermentation plants, on the other hand, can convert the majority of lignocelluloses (hemicellulose, cellulose, and lignin) into usable compounds and this is important for using straw as a biomass feedstock [17]. As a result of higher use of the feedstock, these new improved biomass conversion technologies utilize non-renewable energy sock and produce higher ethanol yield and lead to decline in CO_2 generation than first-generation plants [18, 19]. Furthermore, second-generation plant feedstocks are less expensive (e.g., agricultural waste) and do not raise the behavioral difficulties as potential food resources are not being exploited for energy production. The largest CO_2 reduction is achieved in the second-generation biorefinery with the production of some other useful products. This is accomplished in the Danish bioenergy concept by co-producing hydrogen and methane [20]. An additional key concern is that rich coal supplies are the principal non-renewable energy source in the conversion of biomass to ethanol. As a result, a superior liquid transportation fuel is created from least expensive energy supply, reducing the demand for petroleum imports even more [21]. The environmental impact of ethanol-gasoline blends is determined by the percentage of ethanol in the blend. According to a recent study by Niven, utilizing alcohol as an additive in fuel provides no environmental benefits over using pure gasoline [22]. Presence of 10% ethanol in fuel blend has no effect on GHG emissions or energy efficiency, but it does increase the severity of pollution in groundwater and soil. Although increment of ethanol up to 85% reduces GHG emissions, it results in air pollution in the form of organic and NO_x compounds. [23]. Nevertheless, if 85% ethanol is the aim, existing combustion engines will need to be modified, and various car exhaust catalysts will likely cut NO_x emissions and other organic compounds. Ethanol enhances octane and improvement in combustion have been observed with 10 vol. percent combinations, lowering CO and ozone emissions [21]. Other octane boosters, like as methyl tert-butyl ether and lead, are more toxic to the environment than ethanol (MTBE). Catalytic steam reformation process for the creation of H_2 is another potential future usage of bioethanol to provide a sustainable/green transportation fuel. Many people believe hydrogen will be a future energy source, particularly for transportation and mobile electronics [22]. The majority of hydrogen is now produced by steam reforming natural gas. However, this is not a viable option for reducing GHG emissions in the atmosphere. There are two main reasons why bioethanol to hydrogen conversion is favorable. For transport, combustion of ethanol is projected to have a 20% energy efficiency, whereas using hydrogen fuel cells has a 60% energy efficiency [24]. As a result, even though converting ethanol to hydrogen needs some energy, auto thermal reforming of ethanol may be able to capture more than half of the photosynthetic energy as electricity [25]. Second, removing the water is the most expensive step in using bioethanol directly as a fuel. It is not essential to remove water when steam reforming ethanol to hydrogen, making the process significantly less expensive. utilization of ethanol in a fuel cell is very concerted. Though, the existing direct ethanol fuel cell's efficiency is insufficient [26]. The chemical processes that catalyze steam reforming of bioresources are quite complicated. Computational approaches, on the other hand, can provide specific information on the intermediates and so help pinpoint where catalytic adjustments should be made for large-scale hydrogen production. Beyond 95% of all carbon-containing compounds are derived from fossil fuels. Biomass and biomass-derived compounds are the only available feedstocks for the production of these chemicals. Because the chemical industry is so large, switching to renewable CO_2-neutral resources could result in considerable reductions in GHG emissions. Many regions throughout the world have already established bioethanol production, and many more are in the process of doing so. This indicates that the amount of bioethanol available will almost certainly increase in the next years. With this much bioethanol on hand, there's a lot of potential for partially changing the chemical industry to one that uses renewable resources. Instead of being utilized as a transport fuel, where it is least valued, bioethanol might be used as a resource for other key chemical products. It might be used to make ethene, acetaldehyde, butadiene, and acetic acid, among other things, in addition to production of H_2 via steam reforming or direct usage as a fuel additive. Around 1.4, 120, 7.5, and 8.5 million tons of these chemicals are manufactured annually around the world [27]. Bioethanol is already produced in quantities of roughly 45 million tons per year, indicating that it has the potential to replace a major portion of these petrochemical goods. Bioethanol-derived products must obviously be more valued than the bioethanol itself in order for this to be commercially viable. Furthermore, the conversion operations must be kept as low-cost as possible. Bioethanol costs roughly 250 US dollars per ton, ethylene costs around 700 US dollars per ton, and acetic acid costs around 650 US dollars per ton [5]. With the right catalytic reaction route, it appears feasible to create acetic acid from bioethanol, for example, based on the amounts and production costs of the relevant chemicals. According to a recent proposal, in an aqueous-phase reaction using gold as a catalyst, ethanol is oxidized with air [28]. Furthermore, an examination of the reaction shows that it is both environmentally and economically competitive with the Monsanto process's traditional petrochemical acetic acid production. In terms of CO_2 emissions, the Monsanto process emits roughly 0.7 tons of carbon dioxide for every ton of acetic acid produced, whereas bioethanol production is nearly CO_2 neutral. As a result, producing chemicals from bioethanol can save more CO_2 than using it as a fuel. Because ethylene is produced in large quantities than bioethanol, it is currently impracticable to manufacture all of the required ethene from ethanol. However, with increased bioethanol production capacity and, eventually, rising oil prices, this is expected to change in the future. However, because bioethanol production is expected to be lower in the near future than that of the most important carbonaceous chemicals, it might potentially become a major feedstock for several of them.

9.5 Summary

Petroleum became a vital source of energy and commodity chemicals over the previous century. One long-term objective is to transform biomass into goods that can compete with those made from fossil fuels, with an emphasis on resource availability, sustainability, and supply reliability. By 2025, the US Department of Energy wants to substitute 25% of carbon-containing compounds with chemicals obtained from biomass [15]. Some controllable oxygenated molecules are required to attain this purpose. Ethanol is an excellent choice because of its large-scale synthesis, but other readily available compounds from biomass, such as fructose and glucose can be crucial in achieving this and other goals. In a combined heat and power plant, dry biomass is most energy-efficient where it replaces fossil fuels on a 1:1 basis and achieves energy productivities of over 90%. Moreover, for combined heat and power production, if biomass is used instead of first-generation bioethanol production, a

50% improvement in CO_2 reduction can be realized [29]. It was recently projected that biomass might cover roughly 20% of overall energy use in the European Union in the future, without jeopardizing food security. As a result, biomass cannot meet all of the energy demands [30]. Until far, the primary impetus for ethanol production has been the requirement for transport fuels. Only transport fuels consume about 530 billion liters per year in the United States, whereas in 2005, production of bioethanol was only about 15 billion liters [31]. Bioethanol will almost likely have a good impact on CO_2 emissions in transportation sector, but it is promising to obtain even better environmental results by using biomass. Finding advanced catalytic reaction mechanisms for generating commodity or specialty chemicals from bioethanol, rather than using it to substitute the least priced fossil products, such as fuels, has the greatest economic promise. Thus, the most efficient use of bioethanol may be to incorporate it into valued products that would otherwise be made from petroleum resources and then combust the waste in a thermal power station for the generation of heat. The key argument is that all available resources, both fossil as well as renewable, are inadequate, and it is critical that we carefully analyze how each might be exploited to its maximum potential. This will be a major problem for research in chemistry and advancement in the future.

References

1. British Petroleum Company. (2005). *BP Statistical Review of World Energy June 2005*. British Petroleum Educational Service.
2. Mather, A. A., Garland, G. G., & Stretch, D. D. (2009). Southern African sea levels: Corrections, influences and trends. *African Journal of Marine Science, 31*(2), 145–156.
3. Wigley, T. M. (2005). The climate change commitment. *Science, 307*(5716), 1766–1769.
4. Dale, B. E. (2003). 'Greening'the chemical industry: Research and development priorities for biobased industrial products. *Journal of Chemical Technology & Biotechnology, 78*(10), 1093–1103.
5. Jørgensen, B., Makkee, M., Andersen, J., Dahl, S., & Fehrmann, R. (2008). Catalytic Conversion of Biofuels.
6. Rass-Hansen, J., Falsig, H., Jørgensen, B., & Christensen, C. H. (2007). Bioethanol: fuel or feedstock? *Journal of Chemical Technology & Biotechnology: International Research in Process, Environmental & Clean Technology, 82*(4), 329–333.
7. Kharatyan, S. G. (1978). Microbes as food for humans. *Annual Reviews in Microbiology, 32*(1), 301–327.
8. Schnell, P. G., & Akin, C. (1979). Functional properties of yeast grown on ethyl alcohol. *Journal of the American Oil Chemists' Society, 56*(1), A82–A85.
9. Oki, T., Sayama, Y., Nishimura, Y., & Ozaki, A. (1968). L-Glutamic acid formation by microorganisms from ethanol. *Agricultural and Biological Chemistry, 32*(1), 119–120.
10. Nakayama, K., & Araki, K. (1973). *U.S. Patent No. 3,708,395*. Washington, DC: U.S. Patent and Trademark Office.
11. Hideaki, Y., Kazuoki, O., Terukazu, N., & Yoshihiro, T. (1980). Production of L-amino acids. US Patent 4276380.
12. Chernyavskaya, O. G., Shishkanova, N. V., Il'Chenko, A. P., & Finogenova, T. V. (2000). Synthesis of α-ketoglutaric acid by Yarrowia lipolytica yeast grown on ethanol. *Applied Microbiology and Biotechnology, 53*(2), 152–158.

13. Giudici, P., Gullo, M., & Solieri, L. (2009). Traditional balsamic vinegar. In *Vinegars of the World* (pp. 157–177). Springer, Milano.
14. Strategy, B. (2005). *Key facts and figures*. MEMO/06/05 European Commission, Brussels.
15. Ragauskas, A. J., Williams, C. K., Davison, B. H., Britovsek, G., Cairney, J., Eckert, C. A., Jr Frederick, W. J., Hallett, J. P., Leak, D. J., Liotta, C. L., & Mielenz, J. R. (2006). The path forward for biofuels and biomaterials. *Science, 311*(5760), 484–489.
16. Farrell, A. E., Plevin, R. J., Turner, B. T., Jones, A. D., O'hare, M., & Kammen, D. M. (2006). Ethanol can contribute to energy and environmental goals. *Science, 311*(5760), 506–508.
17. Haagesen, F., Torry-Smith, M., & Ahring, B. K. (2005). Fermentation of biomass to alcohols. *Biofuels for fuel cells: Renewable energy from biomass fermentation* (pp.169–190).
18. Bozbas, K. (2008). Biodiesel as an alternative motor fuel: Production and policies in the European Union. *Renewable and Sustainable Energy Reviews, 12*(2), 542–552.
19. Hammerschlag, R. (2006). Ethanol's energy return on investment: A survey of the literature 1990– present. *Environmental Science & Technology, 40*(6), 1744–1750.
20. Bartacek, J., Zabranska, J., & Lens, P. N. (2007). Developments and constraints in fermentative hydrogen production. *Biofuels, Bioproducts and Biorefining: Innovation for a Sustainable Economy, 1*(3), 201–214.
21. Shapouri, H., Duffield, J. A., & Wang, M. (2003). The energy balance of corn ethanol revisited. *Transactions of the ASAE, 46*(4), 959.
22. Jacobson, M. Z., Colella, W. G., & Golden, D. M. (2005). Cleaning the air and improving health with hydrogen fuel-cell vehicles. *Science, 308*(5730), 1901–1905.
23. Niven, R. K. (2005). Ethanol in gasoline: Environmental impacts and sustainability review article. *Renewable and sustainable energy reviews, 9*(6), 535–555.
24. Larminie, J., Dicks, A., & McDonald, M. S. (2003). *Fuel cell systems explained* (Vol. 2, pp. 207–225). Chichester, UK: J. Wiley.
25. Deluga, G. A., Salge, J. R., Schmidt, L. D., & Verykios, X. E. (2004). Renewable hydrogen from ethanol by autothermal reforming. *Science, 303*(5660), 993–997.
26. Lamy, C., Belgsir, E. M., & Leger, J. M. (2001). Electrocatalytic oxidation of aliphatic alcohols: Application to the direct alcohol fuel cell (DAFC). *Journal of Applied Electrochemistry, 31*(7), 799–809.
27. Weissermel, K., & Arpe, H. J. (2008). *Industrial organic chemistry*. John Wiley & Sons.
28. Christensen, C. H., Jørgensen, B., Rass-Hansen, J., Egeblad, K., Madsen, R., Klitgaard, S. K., Hansen, S. M., Hansen, M. R., Andersen, H. C., & Riisager, A. (2006). Formation of Acetic Acid by Aqueous-Phase Oxidation of Ethanol with Air in the Presence of a Heterogeneous Gold Catalyst. *Angewandte Chemie, 118*(28), 4764–4767.
29. Nielsen, P. H., & Wenzel, H. (2005). *Environmental assessment of ethanol produced from corn starch and used as an alternative to conventional gasoline for car driving*. Technical University of Denmark.
30. Ericsson, K., & Nilsson, L. J. (2006). Assessment of the potential biomass supply in Europe using a resource-focused approach. *Biomass and Bioenergy, 30*(1), 1–15.
31. Editorial. (2006). Bioethanol needs biotech now. *Nature Biotechnol, 24*, 725.

Chapter 10
Bioethanol and Biohydrogen Production from Agricultural Waste

Abstract In recent years, rising energy demand, food security concerns, and rising pollution levels have driven the attention to biofuel and bioenergy production. For decades, petroleum-based fossil fuels have been the most popular, particularly for transport. Though, it is expected that within the upcoming 50 years, fossil fuel sources will be exhausted. As a result, the growth of green fuels is required. Biofuels are made from sustainable and biological raw materials, such as bioethanol, biodiesel, and biogas. Because of their worldwide abundance, these are natural sources that can be used indefinitely. Thus, biofuels are a viable substitute for conventional fuels. This chapter deals with the study of production of two important renewable biofuels—biohydrogen and bioethanol by exploiting agricultural waste as substrate.

Keywords Bioethanol · Biohydrogen · Lignocellulosic biomass · Agricultural waste · Greenhouse gas emission

10.1 Introduction

Bioethanol is currently regarded as one of the most favorable and eco-friendly biofuels for transport and other uses [1, 2] Bioethanol is mostly produced from food crops like sugarcane and corn [3, 4]. US with a national bioethanol yield of 15.78 billion gallons in 2019, is the world's largest producer of bioethanol, followed by Brazil (8.57 billion gallons) [5]. The United States and Brazil, are major providers to the worldwide bioethanol need. Corn and sugarcane are used as primary bioethanol crops and bioethanol produced from such crops is considered 1st generation bioethanol. There are many concerns about food shortage, drought, and other agricultural side effects allied with the development of first-generation biofuels [3]. In recent years, there has been a persistent shift toward a complete examination of non-food alternatives [6]. Second-generation biofuels include biofuels and bioethanol derived from non-food crops [3]. Hydrogen is a resourceful fuel that has been used as rocket fuel, automotive fuel, and chemical industry raw materials. Global hydrogen output was 70 million tons per year in 2019 [7]. Coal gasification, partially oxidizing fossil fuels, and other energy-requiring thermochemical techniques are common ways to make the fuel. Although hydrogen combustion does not

N. K. Aggarwal et al., *Bioethanol Production*, Green Chemistry and Sustainable Technology, https://doi.org/10.1007/978-3-031-05091-6_10

pollute the environment, the process of manufacturing it does [8]. Because of its creation from organic wastes via green process of dark fermentation, biohydrogen has emerged as a more sustainable type of hydrogen. Despite the biofuel's ecologically friendly benefits, the difficulty of low yield combined with lower biomass conversion efficacy still limits its ability to meet the large demand for it [9]. After critical observation of the bioethanol and biogas production processes tells that the budget of production is a major factor impacting their economic feasibility. The difficulties surrounding feedstock logistics and substrate inherent qualities play a substantial role in the overall cost of biofuel manufacturing [10]. The development of economically feasible pretreatment technologies and integrated fermentation models have been highlighted as essential strategies for addressing the problems outlined above in order to manufacture cost-competitive biofuels [11–15]. Other initiatives include the creation of low-cost, high-efficiency technology, the production of valuable side products from the stream produced during biofuel production, and the use of combined biofuel production systems [16–21]

To begin, when addressing the difficulty of feedstock logistics for biofuel production, the feedstock choice must be taken into account. Non-food sources, such as agricultural waste, have the potential to significantly increase the profitability of biofuel production. Agricultural wastes can produce a large amount of renewable energy because of their lignocellulosic properties [22]. Starchy agricultural waste, such as potato peels and cassava peels, are often produced in massive quantities every year all over the world. They are a collection of important carbon polymeric compounds which can be used to make bioenergy and biofuels. Through the implementation of suitable solutions aimed at boosting product yield and process feasibility, starchy wastes can be studied for the production of bioethanol. Waste-based integrated biofuel production and lignocellulosic waste have been described for the production of biohydrogen, bioethanol, biodiesel, and/or biogas production. These fuels are said to be made from a mixture of pure LCB substrates and non-food-based agricultural crops [16, 23–25], or a mix of food crops and agricultural wastes [26]. The utilization of random combinations of starch-based agricultural wastes could thereby address the feedstock logistic difficulty in biofuel operations [27].

An inquiry into the possibility of numerous biofuel streams from mixed starch-derived residual wastes in a combined process is another strategy to meet the current issues of biofuel production. Integrated biofuel production allows more than two biofuels to be produced from biomass having high energy storage, boosting the productivity of the manufacturing process. Agronomic waste-to-biofuel operations produce discharges, which are harmful to the environment and soil microbes. These waste streams contain different organic and inorganic chemicals which can be used to make more biofuel products, boosting the biorefinery's profitability.

10.2 Agricultural Biomass

Agricultural wastes are lignocellulosic feedstock with a little quantity of ash and extractives (10.4–40.5 wt. percent hemicellulose, 25.1–44.3 wt. percent cellulose, and 21.7–44.0 wt. percent lignin) [28]. The cellulose proportion of agricultural waste is generally higher than that of hemicellulose and lignin. A simple comparison of reinforced concrete helps to clarify the functional role of each lignocellulosic component. Lignin, cellulose, and hemicellulose serve as "reinforcing agents," "aggregates," and "hydrated cement," respectively, to offer supplementary structural strength, primary structural strength, and to bind cellulose and lignin together [29, 30]. Individual covalent bonds (ester bonds) and hydrogen bonds bind hemicellulose and lignin to cellulose [31]. Branched polysaccharides like hemicellulose made up of multiple monosaccharides (mainly xylose, mannose, arabinose, and glucose) whereas cellulose is a linear carbohydrate polymer made up of only D-glucose. Meanwhile, lignin is very complex heterogeneous aromatic polymer whose structural stiffness is provided by irregularly organized phenylpropane units [32].

10.3 Why Agricultural Waste Should Be Used?

Agriculture was recognized as one of the economic pillars in several nations with cultivable lands, including Nigeria (21.59%), India (15.39%), Indonesia (13.91%), Malaysia (8.41%), China (8.30%), Denmark (8.30%), and Thailand (8.20%) based on national GDP in 2017 [33]. Rice is one of the most important crops in these nations except Denmark, with paddy fields producing large amounts of rice straws, hulls, and brans [34]. Furthermore, the successful expansion of the oil palm agroindustry in Indonesia, Thailand, Malaysia, and Nigeria leads to the simultaneous generation of oil palm trash [35]. Likewise, the best sugarcane producers (India, China, and Thailand) produce a lot of bagasse during the juice extraction process [36]. To date, Indonesia, Thailand, and Malaysia are the world's top producers of natural rubber. As a result, unproductive rubber trees are capable sources of biomass wastes, such as roots tree, barks, branches, trunks, and leaves, which are similar to logging leftovers. Overall, these agricultural operations result in a large amount of agricultural waste (AW) being generated after the harvest of the primary crop products (Table 10.1), with most Asian nations generating roughly 4.0–5.0 kg/capita/month of agricultural waste by 2025. The massive volume of agricultural waste poses a possible environmental threat if it is just dumped on fertile land. The natural degradation of agricultural waste produces greenhouse emissions (CO_2 and CH_4), which contribute to global warming [37]. Because agricultural waste is frequently packed with soluble organic compounds, unfavorable leaching that occurs during rainy seasons may harm waterways. Rather than being disposed of in landfills, agricultural waste should be appropriately discarded by employing a less polluting and effective technology. In spite of its inherent ecological risks, agricultural waste can be used as a renewable supply of

Table 10.1 Agricultural waste generation by different Asian countries

Country	Agri waste produced (kg/capita) per month	Projected production in year 2025
Singapore	5.0	5.0
Indonesia	3.4	4.5
Malaysia	3.7	6.3
Nepal	1.8	2.7
Thailand	2.9	6.8
Bangladesh	1.2	2.7
Japan	5.1	5.9
China	3.6	4.1
Brunei	3.0	4.3
Laos	2.5	4.1
Cambodia	2.3	5.0
Mayanmar	2.0	3.8

LCB for value-added compounds such as fibreboard, bioproducts, and biofuels via thermochemical processes. The surge in population of humans necessitates an instantaneous search for alternative renewable sources, given the non-renewable fossil fuel stock. Because of their high availability, lignocellulosic agricultural waste is expected to be an eco-friendly, renewable, cost-effective resource for generating bioenergy in the face of an impending energy crisis.

10.4 Agricultural Waste for Production of Biohydrogen

Agriculture is easily the most profitable and productive business on the planet, yet it generates a large amount of trash each year, approximately in billion tons and this garbage contains sugars in both polymeric and simple forms that are almost unused [38]. According to the Food and Agriculture Organization (FAO) of United Nations, a substantial amount of vegetables and fruits are wasted during cultivation, transportation, and marketplace [39]. Because of their carbohydrate content, approximately in million tons of unmarketable vegetables can be used as feedstock for dark fermentation. The excellent degradable parts of the LCB matrix which can be exploited to make biohydrogen by anaerobic fora are cellulose and hemicelluloses [40]. Table 10.2 shows the batch technique for producing biohydrogen from various agricultural leftovers.

Pretreatment of biomass can be done in a variety of ways, including mechanical and physical methods, chemical methods, and green methods. Pretreatments can be used as a single or multiple-step process regarding biomass structure [41]. Physical pretreatment is defined as the reduction in size or disintegration of the biomass structure using a physical force. Higher temperatures are usually followed by both acidic

Table 10.2 Pretreatment of different agri-based substrates [47]

Substrate	Inoculum type/Pretreatment	Yield of biohydrogen (per volatile solid)
Sunflower stalks	Anaerobic digestion	2.3 ml H_2 g
Rice straw	Anaerobic sludge	23.8 ml H_2 g
Mixed agro-industrial waste	Anaerobic granular sludge	4554.5 ml H2 /L
Leaves waste	Sewage sludge	37.9 ml/g/ VS
Sugarcane bagasse	Purple nonsulfur bacteria	1.96 mol H_2 mol sugar $^{-1}$
Beet pulp	Seed sludge	90.3 ml H_2 g COD^{-1}
Wheat straw	Mesophilic anaerobically treated sludge	10.51 ml H_2 g
Wheat stalks	Anaerobic digested activated sludge	23 ml H_2 g

and alkaline chemical treatments. Biological pretreatments are capable of utilizing microorganisms in an ambient working environment, although they are less effective in terms of conversion rates and monomer output from composite carbohydrates [42]. The goal of looking for pretreatment processes is to help microorganisms get at the useful sugars in the biomass [43]. Because of their potential to produce useful biohydrogen energy from biomass and organic wastes, anaerobic bacteria-based fermentation is regarded as a more desirable biological pathway for hydrogen synthesis [44]. For H_2 production, various fermentation methods such as direct biophotolysis, mixed sequential dark-photo, and photofermentation were investigated [42]. Water molecules split into H^+ ions and oxygen in the direct biophotolysis process, which uses solar energy. These hydrogen ions are converted to molecular hydrogen by hydrogenase enzymes. H_2 production was investigated using a variety of cyanobacteria and micro- and macroalgal species [45]. O_2 removal is a problem in this process because it inhibits hydrogenase enzyme activity and hence hinders H_2 synthesis [46].

There are two phases to the indirect biophotolysis system. oxygen is expelled with carbon dioxide fixation in the first step, and H_2 is created in the second stage [45]. Direct biophotolysis can be done in a single reactor to produce oxygen and hydrogen in a fluctuating cycle, or in separate reactors such as photo-bioreactors and open ponds [46].

Bioelectro-hydrogenesis is a type of microbial electrolysis in which current is generated by a bio-electrochemical system that creates hydrogen by reducing protons. The electronic separator, anodic chambers, cathodic chambers, and external electrical power source make up the microbial electrolysis cell (MEC) [48]. This technique produces H2 from industrial and domestic wastewater, as well as agro-industrial leftovers comprising cellulose and starch biopolymers. The performance of the MEC is influenced by key parameters like physicochemical transport microbial physiological mechanisms. The most serious issue remains maintaining electrical potential harmony in both the bioanode and biocathode chambers [49]. The gram-positive bacteria were recognized for enhanced production of biohydrogen under dark conditions. Microbes with endospore production and a quick growth rate are a superior choice for commercial applications. The amount of hydrogen a bacterial culture can extract from glucose is affected by the metabolic pathway and the final product [50]

The presence of light is required for photoheterotrophic bacteria to convert organic acids (e.g., butyric, lactic, and acetic) to carbon dioxide and hydrogen under anaerobic circumstances in photofermentation. As a result, these photoheterotrophic anaerobic microbes turn the produced organic acids into H_2 and CO_2 during the acidogenic reaction. In any externally illuminated photo-bioreactor at commercial size, the photofermenter system must be built with adequate light distribution to avoid shade, and a larger surface area to volume ratio is required [51]. Figure 10.2 depicts the whole chemical reactions indulged in the above-mentioned biohydrogen production. In the production of biohydrogen gas, sequential dark and photofermentation proved to be a successful method. Dark fermentation and photofermentation are linked because the waste from dark fermentation provided enough organic acids for photofermentation, resulting in a larger biohydrogen production than the individual fermentation process [51]. Various microorganisms are capable of producing H_2 from any available renewable substrate in a moderate environmental condition, making a biological method more appealing than a conventional technique [52].

10.5 Bioethanol as Sustainable and Eco-friendly Biofuel

No doubt, bioethanol is major renewable fuel to replace the non-renewable petroleum-based fuels. Globally production of bioethanol has increased from 50 million m^3 in year 2007 to 100 million m^3 in year 2012 [53]. There the two nations-Brazil and the United States are fulfilling around 80% requirements by using corn or sugarcane. Today, it is becoming important to replace the food-based feedstock with non-food materials, for instance- sweet sorghum or cassava. Hydration of ethylene through acid catalysis produces industrial ethanol. Fermentation of feedstock is usually done by the yeast Saccharomyces cerevisiae and bacteria *Zygomonas mobilis*. These organisms metabolize sugars present in substrate and produce ethanol along with carbon dioxide. Ethanol produced by this scheme is used either as alcoholic beverage or as transportation fuel. The main reasons to enhance the use of bioethanol are—economically viable, and eco-friendly since there is less or no emission and renewability.

Biologically, bioethanol is made by the fermentation of sugars present in different sources. The utilization of ethanol as an engine fuel started with its utilization in the interior ignition motor designed by Nikolas Otto in 1897 [54]. Alcohol has been used as fuel since the start of the automobile. When oil crisis happened in year 1970, since then ethanol has been established as an alternative fuel. Different initiatives have been taken by many nations to develop biofuel economically from available raw materials. Brazil and the USA have gained success in terms of process development of biofuel. If we talk about first-generation bioethanol, where sugar feedstock is used, different countries rely on different substrates like Brazil use sugarcane, molasses in India, and corn in US. In spite of the fact that bioethanol from 'first generation advancements' is assessed to increment to in excess of 100 billion liters by 2022 [55]. The materials which compete with food, are not sufficient to meet the requirement [56]. These things have driven the attention toward the use of non-food materials such as agricultural residues, paper waste, municipal solid waste, etc. or lignocellulosic biomass often terms as second-generation feedstock, which is renewable organic material present abundantly in the biosphere [57]. Hence, these unique traits of economics, environment, and renewability make bioethanol an ideal biofuel. Since bioethanol can be easily produced from biomass of crops, therefore, offers an opportunity for farmers to generate another source of income. They can cultivate energy crops to fetch income and at the same time, they will be meeting their food needs. Ethanol has special characteristics in that it acts as closed carbon dioxide cycle in the atmosphere because on combustion the CO_2 released is recycled back into plants as plants during the process of photosynthesis, use this to synthesize cellulose. Production of ethanol only relies on renewable energy resources therefore, there is no emission of any particulate matter or harmful gases. Ethanol has 35% oxygen that aids complete burning of fuel and thus reduces gaseous emission that poses hazards to us. The toxicity level of the exhaust emissions from ethanol is much lower than that of petroleum sources [58]. Along these lines, the utilization of even 10% ethanol blending decreases GHG discharges by 12–19% contrasted and customary petroleum products. On combustion, ethanol emits much lower amount of nitrogen oxide, particulate, and sulfate as compared to conventional fuel [59]. US, Brazil, and China are the largest producers of bioethanol all over the world. In year 2009, by using corn, US produced 39.5×10^9 l of ethanol and by using sugarcane, Brazil produced 30×10^9 l. China is engaged extensively in production of bioethanol and becoming the largest producer [60]. In India, the interest in biofuels is growing so as to substitute oil for achieving energy security and promoting agricultural growth. But now India is also actively participating in the production of biofuels to replace the petroleum-based fuels to secure the energy and environment. Different policies have been designed for promotion of biofuel in terms of production and usage, especially on wastelands. [61].

10.6 Physicochemical Characteristics of Ethanol

Bioethanol is a transparent, colorless liquid with having pleasant odor. Its taste varies according to its concentration. Concentrated has a burning taste, while diluted has sweet flavor. It is the second member alcoholic group which contains hydroxyl group. The melting and boiling points of ethanol are -114.1 °C and 78.5 °C, respectively. Its density is 0.789 g/mL at 20 °C temperature. It can form homogeneous mixture with both polar and non-polar solvents. Ethanol is used as organic solvent in various sectors and products like perfumes, paints, lacquer, and explosive industry. On dehydration, ethanol produces ether. Acetaldehyde and further acetic acid can also be produced

Table 10.3 Physicochemical properties of bioethanol

Parameter	Characteristics
Molecular formula	C2H5OH
Molecular mass	46.07 g/mol
Appearance	Colorless liquid
Water solubility	Between –117 and 78 °C
Density	0.789 kg/l
Boiling point	78.5 °C
Freezing point	−117 °C
Ignition temperature	lowest temperature of ignition
Octane number	99

if ethanol is oxidized. The detailed physicochemical properties of bioethanol are shown in Table 10.3. Alcoholic solutions having non-volatile substances are called tinctures and solution containing volatile substances is called spirit [62].

10.7 Ethanol Production from Crop Residues

In developed countries, crop residues mainly straw and stover are widely used and studied for their potential application in energy. Utilization of crop residues for generating energy may provide security of supply, and their use for ethanol production is strongly sustained [63]. Lignocellulosic feedstock which is rich in sugars can also be used to produce the ethanol [64]. Different methods and procedures for production of ethanol from crop residues have been reported in the literature [65–67]. According to the different studies, each type of feedstock requires pretreatment, specific delignification as well as enzymatic hydrolysis, and fermentation. Different physical and chemical pretreatments like grinding, drying, etc. influence the glucose yield and the conditions of pretreatment depend upon the type of feedstock [63]. A study has reviewed different possible methods for production of ethanol by using corn and rice waste [68]. According to this study, production of bioethanol from corn stover through SSF would be the most economically viable and environmentally friendly process [69]. Nonetheless, it is essential to evaluate the use of residual crops as raw materials for production of ethanol, and in that process take alternative applications into consideration. Crop residues have some other applications like increasing the levels of carbon in soil, to limit the soil erosion, increasing water holding capacity of soil, and enhancing fertility of soil [66]. In future, if technologies in pretreatment procedures develop particularly for lignocellulosic biomass then yield of bioethanol will definitely increase.

10.8 Production of Ethanol from Some Promising Agricultural Feedstock (Based on Data)

Bioethanol is a renewable source which can replace fossil fuels in the transportation. Currently, worldwide 28,375 million gallons of ethanol production are reported which has almost doubled in the past 10 years. The main producers of bioethanol are the United States, Brazil, Europe, and China, followed by Canada, Thailand, Argentina, and India. Among these US contributes 58% of total production [70]. Bioethanol has two main advantages over petrol—1. high-octane number which ignites engine early [71], 2. Higher oxygen content due to which there is lower emission of particulate matter and nitrogen oxide [43]. Therefore, blending bioethanol with petrol is preferably good option and could be a safe environmentally, which is usually used as an octane enhancer for petrol [72]. From the three major crop residues, rice straw, wheat straw, and corn stover, the projected global annual bioethanol production is 205, 104, and 58.6 gigaliters (GL) out of 731.3, 354.34, and 128.02 million tons of available biomass, overall [73]. Likewise, production of bioethanol from other crop residues, such as barley (3.7 million tons) and oat straw (11 million tons) is 20.6 and 3.16 GL, respectively [65]. The production potential of feedstock primarily relies on the composition of cellulose and hemicellulose. Whereas, lignin is the main obstacle in biological conversion into ethanol. Other than this, some mineral and biochemical contents, like high ash and silica present in rice straw; alkali and protein present in wheat straw and corn stover, also cause a limitation to the conversion of biomass into ethanol [74]. Hence, initial pretreatment for any the crop residue is necessary for production of bioethanol. For the crop residue like rice straw and wheat straw, alkaline pretreatment is very effective in fractionation of lignin, and removal of silica content. In addition to silica, other compounds like acetyl group and phenolic compounds should also be removed before pretreatment so that generation of toxic compounds can be decreased, therefore yield of ethanol will be increased [75]. An ethanol yield of 61.3% (29.1 g/L) was obtained in the case of rice straw due to alkali and microwave pretreatment, followed by hydrolysis and fermentation with T cellulase from *S. cerevisiae*. On the other hand pretreatment with just alkali (40% 74% theoretical ethanol) and acid (30% theoretical ethanol) results in lower yields [46]. If wheat straw is chosen, then two stages of pretreatment are required—first extraction phase with sodium hydroxide, second acidic treatment, [75]. Yield of ethanol has been increased by applying harsh chemicals like phosphoric acid and hydrogen peroxide [76]. Likewise, in case of corn stover, by employing several chemical-based pretreatment technologies like alkaline, solvent-based, and ammonia a maximum of 19–22% of ethanol production has been generated. Meanwhile, biological pretreatment (using fungi) gives a much lower ethanol yield of 11% [77]. Various microorganisms (yeasts, bacteria, and fungi) have also been studied for enhancement of yield of ethanol during fermentation. For utilization of wheat straw, *S. cerevisiae, Kluyveromyces marxianus native, Pichia stipites,* and recombinant strains are the most studied yeast strains for ethanol fermentation. Among this *S. cerevisiae* is found to be the best strain for ethanol fermentation. Furthermore, some strict anaerobes like Clostridium sp. and Thermoanaerobacter sp. are also reported for ethanol fermentation at higher temperatures. In case of thermophilic bacteria, their low tolerance to ethanol (0.30 g/L) is the main problem [78]. The fermentation potential of several microbes on wheat straw is 11.831.2 g/L of ethanol [78]. Sugarcane bagasse, which is one of the major agro-industry residue produces 51.3 GL of ethanol from 180.73 million tons of available biomass [42]. Various technologies of pretreatment have been tried for sugarcane baggase and maximum 59.1 g/l yield was obtained after using 1.25% H2SO4 [79]. The treated waste of some other agro-industries, like solid waste from the olive mill or the deoiled olive cake was discovered, having an yearly production of 43,108 kg of dry matter estimated worldwide. Extracted ethanol was 3% of the dry olive deoiled cake [80]. Along with this, potential of some other edible deoiled seed cakes like canola, sunflower, sesame, soy, and peanut are also used for the production [81]. Since these deoiled seed cakes contain protein, fiber, and other nutrients. The fiber having cellulose, hemicellulose, and lignin is converted into bioethanol by applying suitable processes of pretreatment, hydrolysis, and fermentation. Likewise, the non-edible seed cake of *P. pinnata* has also been studied for production of bioethanol (0.088%) [82]. Various fruit wastes such as pineapple peel (8.34%), banana peel (7.45%), apple pomace (8.44%), palm oil empty fruit bunch (14.5%), and a mixture of apple and banana (38%) have also been explored for their capacity to produce bioethanol [83].

10.9 Bioconversion of Lignocellulosic Biomass to Bioethanol

Process of conversion of agricultural biomass into bioethanol has several steps like pretreatment, enzymatic hydrolysis, fermentation, and recovery. To obtain high yield bioethanol, it is important that steps should be done carefully and combined together.

10.9.1 Processing and Pretreatments

Pretreatment of feedstock is the main challenge in the bioconversion of agricultural biomass into ethanol. At the time of pretreatment, the matrix of cellulose and lignin bound by hemicellulose should be broken down to decrease the crystallinity of cellulose and increase the fraction of amorphous cellulose, which is the most suitable form for the action of enzymes [84]. For the purpose of pretreatment, different processes have been developed like physical, physicochemical and biological, etc. to enhance the digestibility. These processes are summarized in Fig. 10.1.

Physical pretreatment methods

Lignocellulosic biomass can be physically treated by grinding, chopping, or milling, this will decrease the size of the particle and increase the surface area [85].

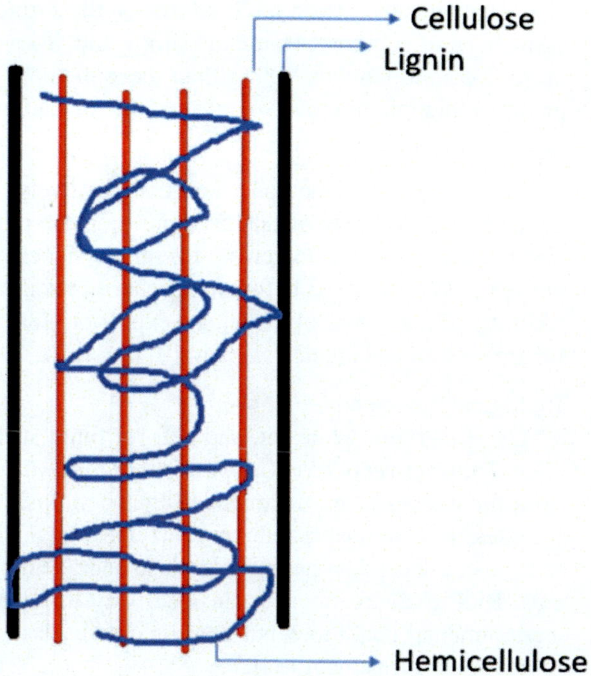

Fig. 10.1 Basic structural arrangement of lignocellulosic biomass

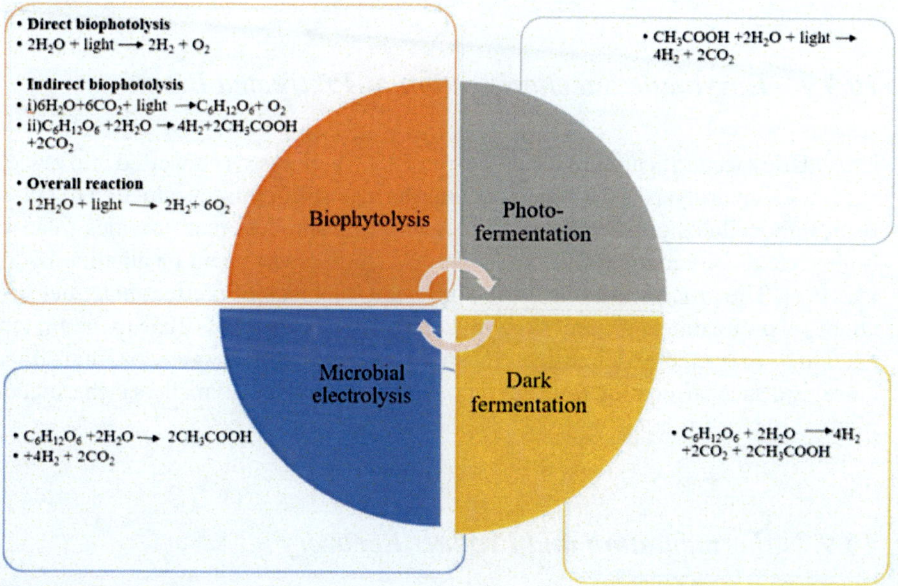

Fig. 10.2 Reactions involved in hydrogen production

This facilitates the action of cellulases on the surface of biomass and improves the conversion of cellulose. Hammer milling and Wiley milling reduce the particle size up to 3–5 mm diameter. Other than these methods like pyrolysis, irradiation with gamma radiation, microwave, infrared, or sonication are also used [86].

Physicochemical methods

Physicochemical methods are comparatively more effective than physical methods. Different chemicals used during these processes are ozone, acids, alkali, peroxide, and organic solvents. Several physicochemical processes are applied for pretreatment of biomass before its saccharification, such as ammonia fiber explosion (AFEX) [87], autohydrolysis (steam explosion) [88], SO_2 steam explosion [89], acid treatment [90] and alkali treatment [91].

Biological treatment

The brown rot, white rot, and soft-rot fungi such as *Phanerochaete chrysosporium, Trametes versicolor, Ceriporiopsis subvermispora,* and *Pleurotus ostreatus* are used for biological pretreatment of lignocellulosic biomass. Polyphenol oxidases, laccases, and quinosine-reducing enzymes, in addition to lignin peroxidases and manganese dependent peroxidases, also degrade lignin by generating aromatic radicals. Biological treatment is an ideal way as it requires low energy and normal environmental conditions but there is drawback of low hydrolysis yield and takes more time for treatment [92].

10.9.2 Enzymatic Saccharification of Pretreated Biomass

Enzymatic saccharification is the process in which cellulose is converted into glucose. Enzymatic hydrolysis is the key to the long-term cost-effective production of ethanol from lignocellulosic materials, as it is a very mild process that provides relatively higher yields and maintenance costs are low compared to acid or alkaline hydrolysis [93]. The process is compatible with many methods of pretreatment, but when chemical pretreatment precedes enzymatic hydrolysis, materials toxic to the enzymes need to be extracted or detoxified. Substrate concentration, enzyme loading, temperature, and saccharification time are factors that affect the enzymatic saccharification method [94].

10.9.3 Fermentation and Product Recovery

By means of cellulolytic enzymes, biomass is converted into fermentable sugars, which are further fermented into bioethanol by application of different microbes. An ideal microbe should be such which can utilize diverse range of substrates, give higher yield, and easy harvesting. Also, that organism should be able to tolerate severe conditions of temperature, ethanol, and inhibitors. Then only after production can be commercially viable.

10.10 Environment Aspects of Bioethanol

Burning of the bioethanol releases CO_2 gas which would be used again by plants. Also, there is no release of CO_2 during conversion of agricultural biomass to bioethanol [95]. This reduces emission of greenhouse gases. Human exposure to bioethanol is harmless. Its degradation rate is also fast in the atmosphere. Use of ethanol along with diesel reduces the octane number, increases the heating value, fractions of aromatics, and changes the temperature of distillation. These properties promote complete ethanol combustion and lower GHG emissions [96]. It also has low photochemical reactivity in the atmosphere and inhibits ozone formation [97].

10.11 Summary

Lignocellulosic feedstock is the key to cost-effective and eco-friendly way to produce bioethanol. Since agricultural waste rich in lignocellulosic material is renewable and present in ample amount. Research on the use of agricultural wastes for the production of second-generation bioethanol has so far produced very promising results worldwide. Several studies have been reported on lab-scale and pilot scale for the production of bioethanol from agricultural waste but still, there is a gap between lab-scale production and industrial production. Therefore, to utilize these cheap, abundantly present, and renewable resources for sustainable production of bioethanol, different hurdles have yet to be overcome. These include-easy and cost-effective process of pretreatment, efficient fermentation by employing ideal microorganisms, and economic product recovery. Considering huge abundance of agricultural feedstock, various efforts are being carried out to make second-generation bioethanol production more sustainable and cost-effective. There seems a huge scope for large-scale production of second-generation biofuel in future.

References

1. Fivga, A., Speranza, L. G., Branco, C. M., Ouadi, M., & Hornung, A. (2019). A review on the current state of the art for the production of advanced liquid biofuels. *Aims Energy, 7*(1), 46–76.
2. Vohra, M., Manwar, J., Manmode, R., Padgilwar, S., & Patil, S. (2014). Bioethanol production: Feedstock and current technologies. *Journal of Environmental Chemical Engineering, 2*(1), 573–584.

3. Ho, D. P., Ngo, H. H., & Guo, W. (2014). A mini review on renewable sources for biofuel. *Bioresource Technology, 169*, 742–749.
4. Zabed, H., Faruq, G., Sahu, J. N., Boyce, A. N., & Ganesan, P. (2016). A comparative study on normal and high sugary corn genotypes for evaluating enzyme consumption during dry-grind ethanol production. *Chemical Engineering Journal, 287*, 691–703.
5. Prasad, S., Kumar, S., Sheetal, K. R., & Venkatramanan, V. (2020). Global climate change and biofuels policy: Indian perspectives. In *Global climate change and environmental policy* (pp. 207–226). Springer, Singapore.
6. De Bhowmick, G., Sarmah, A. K., & Sen, R. (2018). Lignocellulosic biorefinery as a model for sustainable development of biofuels and value-added products. *Bioresource Technology, 247*, 1144–1154.
7. International Energy Agency. *The Future of Hydrogen. 2020.*
8. Aruwajoye, G. S., Kassim, A., Saha, A. K. & Gueguim Kana, E.B. (2020). Prospects for the improvement of bioethanol and biohydrogen production from mixed starch-based agricultural wastes. *Energies, 13*(24), 6609.
9. Sen, B., Aravind, J., Lin, C. Y., Lay, C. H., & Hsieh, P. H. (2019). Biohydrogen production perspectives from organic waste with focus on Asia. In *Biorefinery* (pp. 413–435). Springer, Cham.
10. Oke, M. A., Annuar, M. S. M., & Simarani, K. (2016). Mixed feedstock approach to lignocellulosic ethanol production—prospects and limitations. *BioEnergy Research, 9*(4), 1189–1203.
11. Moodley, P., Sewsynker-Sukai, Y., & Kana, E. G. (2020). Progress in the development of alkali and metal salt catalyzed lignocellulosic pretreatment regimes: Potential for bioethanol production. *Bioresource Technology, 310*, 123372.
12. Aruwajoye, G. S., Faloye, F. D., & Kana, E. G. (2017). Soaking assisted thermal pretreatment of cassava peels wastes for fermentable sugar production: Process modeling and optimization. *Energy Conversion and Management, 150*, 558–566.
13. Sewsynker-Sukai, Y., & Kana, E. G. (2018). Simultaneous saccharification and bioethanol production from corn cobs: Process optimization and kinetic studies. *Bioresource Technology, 262*, 32–41.
14. Aruwajoye, G. S., Sewsynker-Sukai, Y., & Kana, E. G. (2020). Valorization of cassava peels through simultaneous saccharification and ethanol production: Effect of prehydrolysis time, kinetic assessment and preliminary scale up. *Fuel, 278*, 118351.
15. Duque, A., Manzanares, P., González, A., & Ballesteros, M. (2018). Study of the application of alkaline extrusion to the pretreatment of Eucalyptus biomass as first step in a bioethanol production process. *Energies, 11*(11), 2961.
16. Nair, R. B., Kabir, M. M., Lennartsson, P. R., Taherzadeh, M. J., & Horváth, I. S. (2018). Integrated process for ethanol, biogas, and edible filamentous fungi-based animal feed production from dilute phosphoric acid-pretreated wheat straw. *Applied Biochemistry and Biotechnology, 184*(1), 48–62.
17. Dias, M. O., Junqueira, T. L., Rossell, C. E. V., Maciel Filho, R. & Bonomi, A. (2013). Evaluation of process configurations for second generation integrated with first-generation bioethanol production from sugarcane. *Fuel Processing Technology, 109*, 84–89.
18. Boonchuay, P., Techapun, C., Leksawasdi, N., Seesuriyachan, P., Hanmoungjai, P., Watanabe, M., Takenaka, S., & Chaiyaso, T. (2018). An integrated process for xylooligosaccharide and bioethanol production from corncob. *Bioresource Technology, 256*, 399–407.
19. Martinez-Hernandez, E., Sadhukhan, J., & Campbell, G. M. (2013). Integration of bioethanol as an in-process material in biorefineries using mass pinch analysis. *Applied Energy, 104*, 517–526.
20. Yuan, Z., & Wen, Y. (2017). Evaluation of an integrated process to fully utilize bamboo biomass during the production of bioethanol. *Bioresource Technology, 236*, 202–211.
21. Nanssou, P. A. K., Nono, Y. J., & Kapseu, C. (2016). Pretreatment of cassava stems and peelings by thermohydrolysis to enhance hydrolysis yield of cellulose in bioethanol production process. *Renewable Energy, 97*, 252–265.

22. Sivamani, S., Chandrasekaran, A. P., Balajii, M., Shanmugaprakash, M., Hosseini-Bandegharaei, A., & Baskar, R. (2018). Evaluation of the potential of cassava-based residues for biofuels production. *Reviews in Environmental Science and Bio/Technology, 17*(3), 553–570.
23. e Silva, J. O. V., Almeida, M. F., da Conceição Alvim-Ferraz, M., & Dias, J. M. (2018). Integrated production of biodiesel and bioethanol from sweet potato. *Renewable Energy, 124*, 114–120.
24. Prakasham, R. S., Sathish, T., Brahmaiah, P., Rao, C. S., Rao, R. S., & Hobbs, P. J. (2009). Biohydrogen production from renewable agri-waste blend: optimization using mixer design. *International Journal of Hydrogen Energy, 34*(15), 6143–6148.
25. Kim, M. S., & Lee, D. Y. (2010). Fermentative hydrogen production from tofu-processing waste and anaerobic digester sludge using microbial consortium. *Bioresource Technology, 101*(1), S48–S52.
26. Chen, S., Xu, Z., Li, X., Yu, J., Cai, M., & Jin, M. (2018). Integrated bioethanol production from mixtures of corn and corn stover. *Bioresource technology, 258*, 18–25.
27. Mithra, M. G., Jeeva, M. L., Sajeev, M. S., & Padmaja, G. (2018). Comparison of ethanol yield from pretreated lignocellulo-starch biomass under fed-batch SHF or SSF modes. *Heliyon, 4*(10), e00885.
28. Stefanidis, S. D., Kalogiannis, K. G., Iliopoulou, E. F., Michailof, C. M., Pilavachi, P. A., & Lappas, A. A. (2014). A study of lignocellulosic biomass pyrolysis via the pyrolysis of cellulose, hemicellulose and lignin. *Journal of Analytical and Applied Pyrolysis, 105*, 143–150.
29. Doherty, W. O., Mousavioun, P., & Fellows, C. M. (2011). Value-adding to cellulosic ethanol: Lignin polymers. *Industrial Crops and Products, 33*(2), 259–276.
30. Wang, X., Jia, Y., Liu, Z., & Miao, J. (2018). Influence of the lignin content on the properties of poly (lactic acid)/lignin-containing cellulose nanofibrils composite films. *Polymers, 10*(9), 1013.
31. Rowell, R. M. (2005). *Handbook of wood chemistry and wood composites*. CRC Press.
32. Dhyani, V., & Bhaskar, T. (2018). A comprehensive review on the pyrolysis of lignocellulosic biomass. *Renewable Energy, 129*, 695–716.
33. List StatisticsTimes. Of countries by GDP sector composition. The World Factbook (GDP); 2018.
34. Hasan, R., Chong, C. C., Bukhari, S. N., Jusoh, R., & Setiabudi, H. D. (2019). Effective removal of Pb (II) by low-cost fibrous silica KCC-1 synthesized from silica-rich rice husk ash. *Journal of Industrial and Engineering Chemistry, 75*, 262–270.
35. Cheng, Y. W., Lee, Z. S., Chong, C. C., Khan, M. R., Cheng, C. K., Ng, K. H., & Hossain, S. S. (2019). Hydrogen-rich syngas production via steam reforming of palm oil mill effluent (POME)–a thermodynamics analysis. *International Journal of Hydrogen Energy, 44*(37), 20711–20724.
36. Loh, Y. R., Sujan, D., Rahman, M. E., & Das, C. A. (2013). Sugarcane bagasse—The future composite material: A literature review. *Resources, Conservation and Recycling, 75*, 14–22.
37. Wuebbles, D. J., & Hayhoe, K. (2002). Atmospheric methane and global change. *Earth-Science Reviews, 57*(3–4), 177–210.
38. Kumar, A., Gautam, A., & Dutt, D. (2016). Biotechnological transformation of lignocellulosic biomass in to industrial products: An overview. *Advances in Bioscience and Biotechnology, 7*(3), 149–168.
39. FAO, RomeYearbook, F. S. (2013). World food and agriculture. *Food and Agriculture Organization of the United Nations, Rome, 15.*
40. Chatellard, L., Marone, A., Carrère, H., & Trably, E. (2017). Trends and challenges in biohydrogen production from agricultural waste. In *Biohydrogen production: sustainability of current technology and future perspective* (pp. 69–95). Springer, New Delhi.
41. Tu, W. C., & Hallett, J. P. (2019). Recent advances in the pretreatment of lignocellulosic biomass. *Current Opinion in Green and Sustainable Chemistry, 20*, 11–17.
42. Wang, J., & Yin, Y. (2018). Fermentative hydrogen production using various biomass-based materials as feedstock. *Renewable and Sustainable Energy Reviews, 92*, 284–306.

43. Argun, H., Gokfiliz, P., & Karapinar, I. (2017). Biohydrogen production potential of different biomass sources. In *Biohydrogen production: sustainability of current technology and future perspective* (pp. 11–48). Springer, New Delhi.
44. Sen, B., Aravind, J., Kanmani, P., & Lay, C. H. (2016). State of the art and future concept of food waste fermentation to bioenergy. *Renewable and Sustainable Energy Reviews, 53*, 547–557.
45. Holladay, J. D., Hu, J., King, D. L., & Wang, Y. (2009). An overview of hydrogen production technologies. *Catalyzis Today, 139*(4), 244–260.
46. Miandad, R., Rehan, M., Ouda, O. K. M., Khan, M. Z., Shahzad, K., Ismail, I. M. I., & Nizami, A. S. (2017). Waste-to-hydrogen energy in Saudi Arabia: challenges and perspectives. In *Biohydrogen production: Sustainability of current technology and future perspective* (pp. 237–252).
47. Kamaraj, M., Ramachandran, K. K., & Aravind, J. (2020). Biohydrogen production from waste materials: Benefits and challenges. *International Journal of Environmental Science and Technology, 17*(1), 559–576.
48. Miller, A., Singh, L., Wang, L., & Liu, H. (2019). Linking internal resistance with design and operation decisions in microbial electrolysis cells. *Environment International, 126*, 611–618.
49. Liu, H., Grot, S., & Logan, B. E. (2005). Electrochemically assisted microbial production of hydrogen from acetate. *Environmental science & technology, 39*(11), 4317–4320.
50. Gadhe, A., Sonawane, S. S., & Varma, M. N. (2015). Enhanced biohydrogen production from dark fermentation of complex dairy wastewater by sonolysis. *International Journal of Hydrogen Energy, 40*(32), 9942–9951.
51. Zhang, T., Jiang, D., Zhang, H., Jing, Y., Tahir, N., Zhang, Y., & Zhang, Q. (2020). Comparative study on biohydrogen production from corn stover: Photofermentation, dark-fermentation and dark-photo co-fermentation. *International Journal of Hydrogen Energy, 45*(6), 3807–3814.
52. Cai, J., Zhao, Y., Fan, J., Li, F., Feng, C., Guan, Y., Wang, R., & Tang, N. (2019). Photosynthetic bacteria improved hydrogen yield of combined dark-and photofermentation. *Journal of Biotechnology, 302*, 18–25.
53. Kang, Q., Appels, L., Baeyens, J., Dewil, R., & Tan, T. (2014). Energy-efficient production of cassava-based bioethanol. *Advances in Bioscience and Biotechnology, 5*(12), 925.
54. Nag, A. (2008). *Biofuels refining and performance*. McGraw-Hill.
55. Goldemberg, J., & Guardabassi, P. (2009). Are biofuels a feasible option? *Energy Policy, 37*(1), 10–14.
56. Hahn-Hägerdal, B., Galbe, M., Gorwa-Grauslund, M. F., Lidén, G., & Zacchi, G. (2006). Bioethanol–the fuel of tomorrow from the residues of today. *Trends in Biotechnology*, *24*(12), 549–556.
57. Claassen, P. A. M., Van Lier, J. B., Lopez Contreras, A. M., Van Niel, E. W. J., Sijtsma, L., Stams, A. J. M., De Vries, S. S., & Weusthuis, R. A. (1999). utilization of biomass for the supply of energy carriers. *Applied Microbiology and Biotechnology, 52*(6), 741–755.
58. Wyman, C. E., & Hinman, N. D. (1990). Ethanol: Fundamentals of production from renewable feedstocks and use as transportation fuel. *Applied Biochemistry and Biotechnology, 24*, 735–775.
59. Licht, F. O. (2008). World fuel ethanol production. Renewable fuels association.
60. Ivanova, V., Petrova, P., & Hristov, J. (2011). Application in the ethanol fermentation of immobilized yeast cells in matrix of alginate/magnetic nanoparticles, on chitosan-magnetite microparticles and cellulose-coated magnetic nanoparticles. *arXiv preprint* arXiv:1105.0619
61. Ravindranath, N. H., Lakshmi, C. S., Manuvie, R., & Balachandra, P. (2011). Biofuel production and implications for land-use, food production and environment in India. *Energy Policy, 39*(10), 5737–5745.
62. Li, W., Zheng, P., Guo, J., Ji, J., Zhang, M., Zhang, Z., Zhan, E., & Abbas, G. (2014). Characteristics of self-alkalization in high-rate denitrifying automatic circulation (DAC) reactor fed with methanol and sodium acetate. *Bioresource Technology, 154*, 44–50.
63. Soccol, C. R., Faraco, V., Karp, S. G., Vandenberghe, L. P., Thomaz-Soccol, V., Woiciechowski, A. L., & Pandey, A. (2019). Lignocellulosic bioethanol: current status and future perspectives. In *Biofuels: Alternative feedstocks and conversion processes for the production of liquid and gaseous biofuels* (pp. 331–354). Academic Press.

64. Kabel, M. A., Bos, G., Zeevalking, J., Voragen, A. G., & Schols, H. A. (2007). Effect of pretreatment severity on xylan solubility and enzymatic breakdown of the remaining cellulose from wheat straw. *Bioresource Technology, 98*(10), 2034–2042.
65. Kim, S., & Dale, B. E. (2004). Global potential bioethanol production from wasted crops and crop residues. *Biomass and bioenergy, 26*(4), 361–375.
66. Reijnders, L. (2008). Ethanol production from crop residues and soil organic carbon. *Resources, Conservation and Recycling, 52*(4), 653–658.
67. Li, X., Mupondwa, E., Panigrahi, S., Tabil, L., Sokhansanj, S., & Stumborg, M. (2012). A review of agricultural crop residue supply in Canada for cellulosic ethanol production. *Renewable and Sustainable Energy Reviews, 16*(5), 2954–2965.
68. Arvanitoyannis, I. S., & Tserkezou, P. (2008). Corn and rice waste: A comparative and critical presentation of methods and current and potential uses of treated waste. *International Journal of Food Science & Technology, 43*(6), 958–988.
69. Champagne, P. (2007). Feasibility of producing bioethanol from waste residues: A Canadian perspective: Feasibility of producing bioethanol from waste residues in Canada. *Resources, Conservation and Recycling, 50*(3), 211–230.
70. Renewable fuel association. (2017). http://www.ethanolrfa.org/resources/industry/statistics/
71. Balat, M. (2007). Global biofuel processing and production trends. *Energy Exploration & Exploitation, 25*(3), 195–218.
72. Green, K. R., & Lowenbach, W. A. (2001). MTBE contamination: Environmental, legal, and public policy challengesguest editorial. *Environmental Forensics, 2*(1), 3–6.
73. Saini, J. K., Saini, R., & Tewari, L. (2015). Lignocellulosic agriculture wastes as biomass feedstocks for second-generation bioethanol production: concepts and recent developments. *3 Biotech, 5*(4), 337–353.
74. Binod, P., Sindhu, R., Singhania, R. R., Vikram, S., Devi, L., Nagalakshmi, S., Kurien, N., Sukumaran, R. K., & Pandey, A. (2010). Bioethanol production from rice straw: An overview. *Bioresource Technology, 101*(13), 4767–4774.
75. Yuan, W., Gong, Z., Wang, G., Zhou, W., Liu, Y., Wang, X., & Zhao, M. (2018). Alkaline organosolv pretreatment of corn stover for enhancing the enzymatic digestibility. *Bioresource Technology, 265*, 464–470.
76. Qiu, J., Ma, L., Shen, F., Yang, G., Zhang, Y., Deng, S., Zhang, J., Zeng, Y., & Hu, Y. (2017). Pretreating wheat straw by phosphoric acid plus hydrogen peroxide for enzymatic saccharification and ethanol production at high solid loading. *Bioresource Technology, 238*, 174–181.
77. Zhao, Y., Damgaard, A., & Christensen, T. H. (2018). Bioethanol from corn stover–a review and technical assessment of alternative biotechnologies. *Progress in Energy and Combustion Science, 67*, 275–291.
78. Talebnia, F., Karakashev, D., & Angelidaki, I. (2010). Production of bioethanol from wheat straw: An overview on pretreatment, hydrolysis and fermentation. *Bioresource Technology, 101*(13), 4744–4753.
79. Cardona, C. A., Quintero, J. A., & Paz, I. C. (2010). Production of bioethanol from sugarcane bagasse: Status and perspectives. *Bioresource Technology, 101*(13), 4754–4766.
80. Tayeh, H. A., Najami, N., Dosoretz, C., Tafesh, A., & Azaizeh, H. (2014). Potential of bioethanol production from olive mill solid wastes. *Bioresource Technology, 152*, 24–30.
81. Balan, V., Rogers, C. A., Chundawat, S. P., da Costa Sousa, L., Slininger, P. J., Gupta, R., & Dale, B. E. (2009). Conversion of extracted oil cake fibers into bioethanol including DDGS, canola, sunflower, sesame, soy, and peanut for integrated biodiesel processing. *Journal of the American Oil Chemists' Society, 86*(2), 157–165.
82. Muktham, R., Ball, A. S., Bhargava, S. K., & Bankupalli, S. (2016). Bioethanol production from non-edible de-oiled Pongamia pinnata seed residue-optimization of acid hydrolysis followed by fermentation. *Industrial Crops and Products, 94*, 490–497.
83. Gupta, A., & Verma, J. P. (2015). Sustainable bioethanol production from agro-residues: A review. *Renewable and Sustainable Energy Reviews, 41*, 550–567.

84. Lynd, L. R. (1996). Overview and evaluation of fuel ethanol from cellulosic biomass: Technology, economics, the environment, and policy. *Annual Review of Energy and the Environment, 21*(1), 403–465.
85. Wyman, C. E., Dale, B. E., Elander, R. T., Holtzapple, M., Ladisch, M. R., & Lee, Y. Y. (2005). Comparative sugar recovery data from laboratory scale application of leading pretreatment technologies to corn stover. *Bioresource Technology, 96*(18), 2026–2032.
86. Mosier, N., Wyman, C., Dale, B., Elander, R., Lee, Y. Y., Holtzapple, M., & Ladisch, M. (2005). Features of promising technologies for pretreatment of lignocellulosic biomass. *Bioresource Technology, 96*(6), 673–686.
87. Sun, Y., & Cheng, J. (2002). Hydrolysis of lignocellulosic materials for ethanol production: A review. *Bioresource Technology, 83*(1), 1–11.
88. Grous, W. R., Converse, A. O., & Grethlein, H. E. (1986). Effect of steam explosion pretreatment on pore size and enzymatic hydrolysis of poplar. *Enzyme and Microbial Technology, 8*(5), 274–280.
89. Sipos, B., Réczey, J., Somorai, Z., Kádár, Z., Dienes, D., & Réczey, K. (2009). Sweet sorghum as feedstock for ethanol production: Enzymatic hydrolysis of steam-pretreated bagasse. *Applied Biochemistry and Biotechnology, 153*(1), 151–162.
90. Gámez, S., González-Cabriales, J. J., Ramírez, J. A., Garrote, G., & Vázquez, M. (2006). Study of the hydrolysis of sugar cane bagasse using phosphoric acid. *Journal of food engineering, 74*(1), 78–88.
91. Kaar, W. E., & Holtzapple, M. T. (2000). Using lime pretreatment to facilitate the enzymic hydrolysis of corn stover. *Biomass and Bioenergy, 18*(3), 189–199.
92. Brown, R. C., & Brown, T. R. (2013). *Biorenewable resources: Engineering new products from agriculture*. John Wiley & Sons.
93. Kuhad, R. C., Singh, A., & Eriksson, K. E. L. (1997). Microorganisms and enzymes involved in the degradation of plant fiber cell walls. *Biotechnology in the pulp and paper industry* (pp. 45–125).
94. Tucker, M. P., Kim, K. H., Newman, M. M., & Nguyen, Q. A. (2003). Effects of temperature and moisture on dilute-acid steam explosion pretreatment of corn stover and cellulase enzyme digestibility. In *Biotechnology for fuels and chemicals* (pp. 165–177). Humana Press, Totowa, NJ.
95. Marszałek, J., & Kamiński, W. (2008). Environmental impact of bioethanol production. *Proceedings of ECOpole, 2*(1), 65–70.
96. Ch, A. K., Chan, E. S., Rudravaram, R., Narasu, M. L., Rao, L. V., & Ravindra, P. (2007). Economics and environmental impact of bioethanol production technologies: An appraisal. *Biotechnology and Molecular Biology Reviews, 2*(1), 14–32.
97. Wyman, C. E. (1994). Ethanol from lignocellulosic biomass: Technology, economics, and opportunities. *Bioresource Technology, 50*(1), 3–15.

Chapter 11
Ethanogenic Bacteria: Present Status for Bioethanol Production

Abstract Despite growing environmental challenges, large volumes of fossil fuels are consumed every day. Biofuels created from various types of biomass are being used worldwide to preserve the environment and build a sustainable society. Although bioethanol has been widely produced, food crops such as corn and sugar cane are frequently used as substrates. Instead of grain biomass, lignocellulosic feedstock should be employed for bioethanol production to produce a sustainable energy supply. In addition to this, different kinds of microorganisms should be explored for better fermentation and production. Ethanogens are being employed for production of bioethanol. This chapter mainly focuses on the different type of microorganisms and different engineering practices to enhance the rate of ethanol production.

Keywords Ethanogen · Fermentation · Microorganisms · Bioethanol · Genetic engineering

11.1 Introduction

The demand for refined fossil fuels has become overburdened as a result of rapid global industrialization. This need, combined with the rising expense of refined fossil fuels, as well as their significant impact on greenhouse gas emissions and global warming, poses serious socioeconomic issues. Due to declining fossil fuel supplies and improved biofuel production methods, there is an existing global interest in discovering alternate sources of energy to improve human welfare. It is crucial to understand the molecular mechanisms by which these microbes adapt to bioethanol production in the most cost-effective and environmentally friendly way possible [1]. As a result, the development of environmentally sustainable and economical energy sources is required. Bioethanol is one such promising environmentally friendly, cost-effective, and long-term option. Bioethanol has a number of advantages over fossil fuels such as gasoline, diesel, and kerosene, one of which is that it emits significantly fewer harmful gases [2]. These lignocellulosic inhibitory chemicals may have developed to tolerate or fight ethanologenic bacteria. Ethanologenic bacteria and fungi are both used in microbial fermentation for ethanol generation. Microbial acceptance of organic solvents, chemicals, and other bioproducts, produced during the production

N. K. Aggarwal et al., *Bioethanol Production*, Green Chemistry and Sustainable Technology, https://doi.org/10.1007/978-3-031-05091-6_11

of biorenewable fuels [3, 4]. The utilization of genetically engineered ethanologenic microbes aids in ethanol production and the use of residual biomass after fermentation due to improved sugar conversion, leaving a lesser amount of residual biomass [5], and a variety of microbes are commonly employed to synthesize bioethanol [6] from various sugars extracted from agro-industrial waste and photosynthetic microbial biomass [7]. Bioethanol, being a renewable and clean fuel with significant environmental advantages, is now a capable biofuel with the potential to offer a long-term alternative to existing oil-based fuels. Scientists are concentrating their emphasis on the use of residual wastes and byproducts as substrates for the production of bioethanol in order to reduce the competition between fuels and food production. In any school cafeteria, leftover food waste is produced in large quantities, and managing them might be difficult. This contains considerable amounts of sugars (both soluble and insoluble) and can be utilized to make ethanol as a raw material. The chapter mainly focused on different types of microbes used for ethanolic fermentation [8]. In the production of bioethanol or biogas, ethanogenic microorganisms work as a collective 'mini-factory.' The ethanologenic bacteria must meet a strict isolation condition, which involves the use of a low-cost growth medium. It should be able to quickly transform the substrate into the product and recover the product from the culture medium. Depending on the type of fuel, these microorganisms have different fermentation characteristics. There are not any wild strain ethanologens that can ferment all of the sugars generated by hydrolysis in lignocellulosic material. This big obstacle encourages the use of metabolic engineering to create enhanced ethanologens by combining beneficial features from diverse microorganisms. *Escherichia coli* and *Zymomonas mobilis* are the most commonly investigated ethanologens for plant biomass conversion [9]. Many prospective candidates for bioethanol production at the fermentation stage include *Saccharomyces cerevisiae, Kluyveromyces marxianus, Dekkera bruxellensis,* and *Scheffersomyces stipitis* which are extensively employed microorganisms in starch-based bioethanol synthesis [10]. The conversion of lignocellulosic sugars to ethanol is a promising process that requires the assistance of an ethanologenic microbe that can create ethanol with higher yields and productivity while also being tolerant of ethanol and resistant to inhibitory compounds. In the near future, recombinant DNA technology will be used to produce ethanol from LCB at commercial scale by introducing foreign genes and eliminating competitive pathways. *Klebsiella pneumoniae, K. oxytoca, Lactobacillus* strains (*L. pentoaceticus*), *L. casei, L. xylosus, Leuconostoc mesenteroides, Clostridium acetobutilicum, C. thermocellum, Z. mobilis,* and *L. pentosus* are among the bacteria that ferment carbohydrates to ethanol [11, 12]. For an economically effective conversion from LCB to ethanol, ethanologenic bacteria that can ferment additional sugars like mannose, arabinose, xylose, and galactose are necessary.

11.2 Current State of Bacteria Developed for Ethanol Production

11.2.1 E. coli *Genetic Engineering for Ethanol Production*

E. coli is a promising host for converting LCB to ethanol because it can ferment a broad spectrum of sugars, like xylose, arabinose, glucose, galactose, and mannose, and it does not require sophisticated growth factors [13]. The deletion of competing schemes has been exploited to increase sugar utilization, volumetric ethanol yield, and ethanol and inhibitor tolerance using plasmids or foreign genes cloned into the *E. coli* genome or incorporation of foreign genes [14]. Each *E. coli* strain relied on a plasmid to express the *adh* and *pdc* genes. In the absence of antibiotics, plasmids harboring the *adh* and *pdc* genes were not easily lost, but these strains were still too unstable genetically for use in commercial fermentations. The following steps were taken: I chromosomal mixing of genes encoding *pdc* or *adhII* from *Z. mobilis* under the control of the pyruvate formate lyase promoter, (ii) removal of one fumarate reductase subunit (to avoid the synthesis of succinate under conditions of fermentation); (iii) and selection of variants (by adaptive evolution in plates with a high *E. coli LY01* [15], *E. coli FBR5* (bioethanol synthesis from xylose sugar in batch fermentors) [16], and *E. coli K011* (1st report of a metabolically engineered bacteria for the synthesis of ethanol) are among the strains being metabolically engineered for the production of fuel ethanol [17], *E. coli MG1655* was firstly described on cellobiose and xylose co-metabolism [18], *E. coli MM160* pentose and hexose sugars from phosphoric acid pretreated sugarcane bagasse were co-fermented (SScF) to bioethanol in a single container [19] and *E. coli MM170* [20]. The ability of any microorganism to convert renewable raw materials into fuel with high productivity at a low cost without remaining harmful to the organism is critical to its success in industrial fuel production. *E. coli* has emerged as the best microbe for producing biofuels from renewable energy sources due to the availability of molecular and genetic tools for engineering existing inherent pathways or creating a synthetic new pathway.

11.2.2 Engineering in Z. mobilis *for Bioethanol Production*

Z. mobilis is the first recombinant strain that can ferment xylose, and it has a better ethanol production and other benefits as an ethanologen. However, because it exclusively ferments glucose, fructose, and sucrose, it is not well suited for biomass conversion. The first recombinant strain was designed to ferment xylose, and four *E. coli* genes were used to accomplish so: xylose isomerase (xylA), transketolase (tktA), xylulose kinase (xylB), and transaldolase (talB) [21]. The enzymes xylose isomerase and xylulose kinase convert xylose to xylulose-5-phosphate, a pentose that can enter the pentose phosphate cycle. Transketolase and transaldolase then convert xylulose-5-phosphate to ED pathway intermediates. Using strong constitutive promoters of either enolase or glyceraldhyde-3-phosphate dehydrogenase from *Z. mobilis*, the genes were cloned into a plasmid (pZB5). The converted CP4 with pZB5 strain thrived on xylose and produced 86 percent of the expected ethanol yield. This strain fermented both glucose and xylose at the same time. The native glucose permease is required for xylose uptake; *Z. mobilis* lacks active sugar transport mechanisms [22]. The *ATCC 39,676 Z. mobilis* strain, was changed with the genes coding for five enzymes (1) l-arabinose isomerase *(araA)*, (2) l-ribulose kinase *(araB)*, (3) l-ribulose- 5-phosphate-4-epimerase *(araD)*, (4) transketolase *(tktA)*, and (5) transaldolase *(talB)*, successfully fermented arabinose to ethanol [23]. *ATCC 39,767*, a strain of *Z. mobilis* bearing pZB5 having cellulase genes, was able to convert cellulose feedstock to ethanol efficiently, and the researchers were able to improve their *Z. mobilis* strains. The novel stain (*AX101, ATCC 39,676*) ferments both arbinose and xylose and carries seven heterologous genes. When *AX101* was used to ferment a sugar combination, it fermented every molecule of glucose and xylose [24].

11.2.3 Engineered Klebsilla *for Bioethanol Production*

Klebsiella oxytoca can thrive at pH as low as 5.0 and temperatures as high as 35 °C and has been observed cultivating in pulp and paper streams as well as around other sources of wood. Cellobiose, cellotriose, pentoses, and hexoses are among the sugars used by *K. oxytoca* [25]. *K. oxytoca BW21* to produce more and use urea as the only source of nitrogen, allowing cellulose fermentation to take place without the utilization of xylosidases and -glucosidadases [26]. This strain, which contains two endoglucanases from *Erwiniachrys anthemi* as well as genes for ethanol production from *Z. mobilis*, is in high call for bioethanol production. Designed specifically for use in the SSF process, they rapidly synthesize ethanol from both glucose and cellobiose at high yields [27]. The use of cellulose, hemicellulose and starch by anaerobic thermophiles has been demonstrated. *Clostridium thermocellum, C. raminisolvens, and C. stercorarium* are among these species [28]. *C. thermohydrosulphuricum, C. themosaccharolyticus,* and *Thermoanaerobacter ethanolicus* are noncellulolytic thermophiles that use xylose and pentoses at comparable quantities to hexose. The production of ethanol by thermophilic organisms has several advantages: (1) minimal danger of contamination, (2) enhanced diffusion rates, (3) improved mass transfer, (4) lower bioreactor running costs, and (5) increased substrate solubility [29]. *Thermo. saccharolyticum JW/SL-YS485*, a thermophilic anaerobic bacterium that can ferment hemicellulose and xylan polymers directly, thrives at a temperature range of 45–65 °C and a pH range of 4.0–6.5. It also ferments cellobiose, glucose, xylose, mannose, galactose, and arabinose, which are all found in cellulosic biomass [30]. *Clostridium thermocellum* has been described to directly ferment cellulosic biomass without the need for pretreatment, however, due to its low ethanol tolerance, it has not been commercially utilized. *Clostridium thermocellum* strains *SS21* and *SS22* were isolated and found to produce excellent yields of ethanol while also being resistant to different solvents including acetic acid [31]. *Clostridium phytofermentans* mesophilic, a Gram-negative, straight rod, motile, cellulolytic, spherical terminal spore producer that may use cellulose, polygalactouronic acid, pectin, starch, xylan, arabinose, cellobiose, glucose, lactose, maltose, galactose, fructose mannobiose, fructose, galact Ethanol, acetate, CO_2, and H_2 are the principal end-products of this fermentation, while formate and lactate are minor products [32].

11.2.4 Engineering Bacillus *for Production of Bioethanol*

Recombinant cellulolytic *Bacillus* has a number of advantages, including (1) rapid growth with low nutrient requirements (2) high protein secretion capability (3) industrial safety, and the ability to use soluble pentose, hexose sugars such as glucose, xylose, mannose, cellobiose, and others, as well as tolerance to very high salt and solvent concentrations. *B. subtilis* is thought to be an ideal host for the production of secretary protein, which can help with cellulose enzymatic hydrolysis [33]. *B. subtilis MBG 874* is a new stain. Ethanol or lactate production in a single step from cellulose or pretreated biomass produced by recombinant *B. subtilis* [34] (Tables 11.1 and 11.2).

11.3 Different Methods to Improve Production of Bioethanol

11.3.1 Engineering for Utilization of Lignocellulosic Feedstock

Microorganisms create numerous enzymes that digest cellulose and hemicellulose into fermentable sugars by both cellulolytic and hemicellulolytic activity, which is employed for biofuel generation because of its carbon supply. Endoglucoconase (cleave cellulose at random locations, producing oligosaccharides of various lengths), exogluconase (acts on the reducing or nonreducing end of polymer, producing free form of glucose or cellobiose as primary products), and b-glucosidase are three types of cellulolytic enzymes (hydrolyzes cellobiose into glucose). B-mannanases, xylanases, b-xylosidases, arabinofuranosidases, and b-xylosidases are all hemicellulases. Most microorganisms cannot hydrolyze and use LCB, therefore metabolic engineering could be a good way to boost their productivity [42]. Cellulase enzymes work together to digest complicated cellulose into simpler sugars, which microorganisms use for growth and fermentation [43].

Table 11.1 Typical characteristics of Representative Microorganisms for Biofuel Production

Strain	Pros	Cons	References
E. coli	Utilization of sugar	Unaffected to environmental stress, also lower ethanol as well as butanol tolerance	[35]
Z. mobilis	Gives higher yield of ethanol and productivity; and high tolerance to ethanol	Cannot ferment pentoses	[36]
Clostridium phytofermentans (ethanol), *Clostridium acetobutylicum* (butanol)	Saccharify cellulose and hemicellulose, can ferment both hexoses and pentoses	Slow rate of growth, low productivity, and sensitive to infection of several bacteriophages	[32]
Clostridium thermocellum	Thermophilic anaerobe that grows faster on cellulose, both cellulolytic and ethanologenic bacteria, can digest homocellulose and directly ferment hexoses to ethanol and organic acids, no need for external enzyme addition	Cannot ferment on pentose, branched fermentation schemes lead to accumulation of acetate and lactate as byproducts, lower yield of ethanol production, and low tolerance to ethanol	[37]

11.3.2 Genetic Engineering

It is a potential method for creating inhibitor-tolerant bacteria by modifying their genomes. Furfural is highly harmful since it causes reactive oxygen species to build up inside cells, affecting the cytoskeleton, membrane of vacuole, mitochondria, and the pentose phosphate pathway [44].

11.3.3 Selection of Microbes and Cultural Conditions

Because each microorganism has a distinct capacity to tolerate and break down various inhibitors, screening is necessary to select the microorganisms to target in the experiment. Inhibitor effects can also be mitigated by carefully selecting cultural conditions. For biofuel production, simultaneous saccharification and fermentation (SSF) and consolidated bioprocess (CBP) fermentation processes may be useful [45].

Table 11.2 Some recent studies associated with the use of modified microbes for ethanolic fermentation

Microbe (strain)	Modification method	Main result	References
K. pneumoniae GEM167	c-irradiation mutant Then overexpression of *Z. mobilis pdc* and *adhII* genes coding for aldehyde dehydrogenase (Adh) and pyruvate decarboxylase (Pdc)	The assessed cost of ethanol produced from glycerol, in view of both demand for the feedstock and operational costs, is 40% less than the calculated cost of production from corn-based sugars	[38]
C. acetobutylicum ATCC 824	Incorporation of glcG disruption and genetically overexpression of the xylose scheme	Enhanced D-xylose and L-arabinose consumption with better results than the results associated with the wild-type strain	[39]
Z. mobilis	A portion conferring cellulase activity from which threefold increase was seen than the amount obtained with the wild type strain, i.e., *Enterobacter cloacae* which were cloned in *E. coli*, and then the cellulose gene was incorporated into *Z. mobilis*	Using carboxymethyl cellulose and 4% NaOH pretreated bagasse as substrates	[40]
B. subtilis JY123	Expression of arabinose:H + symporter, AraE protein from *B. subtilis*	Efficient transport of D-xylose	[41]

11.3.4 Engineering for Biofuel Tolerance

Microorganisms possess biosynthetic pathways in biofuel generation, hence microbes with improved tolerance for the end product must be created. To reduce the biofuel's inherent toxicity, researchers are concentrating on the microbe's internal structure, such as heterologous expression of efflux pumps (which expel toxins from the cell) or heat shock proteins, membrane protein modification, stress response gene activation, and its operon, which can be cloned to enhance the tolerance [3].

11.4 Current Status of Production and Future Perspective

Despite ongoing technological advancements, commercial-scale production of second-generation biofuels remains a significant problem. Ethanol is the most common biofuel produced around the world, with sugarcane in Brazil and India, corn in the US, and wheat and barley in Europe as the primary feedstocks. In comparison to sugarcane fermentation, ethanol fermentation utilizing maize produces a higher concentration of ethanol [46]. Only a few nations have set up biodiesel manufacturing plants using LCB, including Indonesia and Malaysia (palm oil), India (Jatropha oil), Europe and Canada (rapeseed oil), and the United States (soybean oil) [47]. Biofuels production from LCB is now at the laboratory scale, with just a few countries able to put up commercial units employing lignocellulosic material as a raw material (Brazil, the U.S., and Italy). Only a few microbes have the ability to produce biofuel, and more than 99 percent of the microbial flora has yet to be discovered. There is a need to discover new microbes, as well as engineer existing microbes, for use of a wide variety of carbon sources, tolerance to biomass-based inhibitory compounds, and biofuel production. Coproduction of other valuable byproducts could be a mitigating strategy to improve the economics of biofuel generation [48]. Government policies should be designed to encourage the development and commercialization of second-generation biofuels. More money for biofuel research, as well as subsidies to create infrastructure and launch new businesses, is required. To bring the efforts of many researchers together, international collaboration is essential. This might help investors save money and reduce risk. The production of second-generation biofuels is still in the early stages of development, with the goal of bringing it to the next level.

11.5 Summary

Because it can produce both hexose and pentose sugars during the fermentation process, lignocellulosic biomass has gained favor as a possible future for biofuel production. We can change both microbes and plants with the use of biotechnology and recombinant DNA technologies to get the required products, which can improve biofuel production. The biggest problem with this biomass is lignin, but we can get rid of it with the help of pretreatments, and as research and development progress, the world will obtain more biofuel that does not emit CO2, which is the most favorable aspect of using biofuel. The planet will become greener, and pollution risks will be reduced. We know that fossil fuel sources will be depleted one day, so we should concentrate on lignocellulosic biomass for a stable and renewable energy source. Because of the low demand and great availability, feedstocks and biomass will become more popular. The fundamental pillars for the alteration of biomass, bacteria, and fungus for more increased biofuel production are recombinant DNA technology. For ethanol production, lignocellulosic biomass is an effective raw source. There are various technological obstacles in the lignocellulosic biomass-to-ethanol conversion process, with the selection of suitable microbe being the most difficult. Because inhibitors occur at the time of pretreatment and hydrolysis steps, their applicability is limited, and ethanologenic organisms that can tolerate these inhibitors are required. Exploration of wild and harsh environmental niches could lead to the discovery of novel ethanologenic bacteria that are more resistant to inhibitors. The hunt for new ethanologenic microbes, as well as improvements in fermentation procedures, may aid in the development of cost-effective lignocellulosic ethanol production.

References

1. Harris, J. M., Roach, B., & Environmental, J. M. H. (2007). *The economics of global climate change*. Global Development and Environment Institute Tufts University.
2. Hill, J., Nelson, E., Tilman, D., Polasky, S., & Tiffany, D. (2006). Environmental, economic, and energetic costs and benefits of biodiesel and ethanol biofuels. *Proceedings of the National Academy of sciences, 103*(30), 11206–11210.
3. Dunlop, M. J. (2011). Engineering microbes for tolerance to next-generation biofuels. *Biotechnology for Biofuels, 4*(1), 1–9.
4. Dunlop, M. J., Dossani, Z. Y., Szmidt, H. L., Chu, H. C., Lee, T. S., Keasling, J. D., Hadi, M. Z., & Mukhopadhyay, A. (2011). Engineering microbial biofuel tolerance and export using efflux pumps. *Molecular Systems Biology, 7*(1), 487.
5. Lopes, T. F., Cabanas, C., Silva, A., Fonseca, D., Santos, E., Guerra, L. T., Sheahan, C., Reis, A., & Gírio, F. (2019). Process simulation and techno-economic assessment for direct production of advanced bioethanol using a genetically modified Synechocystis sp. *Bioresource Technology Reports, 6*, 113–122.
6. Singh, N., Puri, M., Tuli, D. K., Gupta, R. P., Barrow, C. J., & Mathur, A. S. (2018). Bioethanol production potential of a novel thermophilic isolate Thermoanaerobacter sp. DBT-IOC-X2 isolated from Chumathang hot spring. *Biomass and Bioenergy, 116*, 122–130.
7. Shankar, K., Kulkarni, N. S., Jayalakshmi, S. K., & Sreeramulu, K. (2019). Saccharification of the pretreated husks of corn, peanut and coffee cherry by the lignocellulolytic enzymes secreted by Sphingobacterium sp. ksn for the production of bioethanol. *Biomass and Bioenergy, 127*, 105298.
8. Tajarudin, H. A., Azmi, M. S., Makhtar, M. M. Z., Othman, M. F., & Ahmad, M. I. (2020). Second-generation bioethanol: Advancement of ethanologenic microorganisms toward industrial production. In *Green Engineering for Campus Sustainability* (pp. 61–79). Springer, Singapore.
9. Taylor, M. P., Mulako, I., Tuffin, M., & Cowan, D. (2012). Understanding physiological responses to pre-treatment inhibitors in ethanologenic fermentations. *Biotechnology Journal, 7*(9), 1169–1181.
10. de Souza Liberal, A. T., Basílio, A. C. M., do Monte Resende, A., Brasileiro, B. T. V., Da Silva-Filho, E. A., De Morais, J. O. F., Simoes, D. A., & de Morais Jr, M. A. (2007). Identification of Dekkera bruxellensis as a major contaminant yeast in continuous fuel ethanol fermentation. *Journal of Applied Microbiology, 102*(2), 538–547.
11. Balat, M., & Balat, H. (2008). A critical review of bio-diesel as a vehicular fuel. *Energy Conversion and Management, 49*(10), 2727–2741.
12. He, M. X., Wu, B., Qin, H., Ruan, Z. Y., Tan, F. R., Wang, J. L., Shui, Z. X., Dai, L. C., Zhu, Q. L., Pan, K., & Tang, X. Y. (2014). Zymomonas mobilis: A novel platform for future biorefineries. *Biotechnology for Biofuels, 7*(1), 1–15.
13. Alterthum, F., & Ingram, L. O. (1989). Efficient ethanol production from glucose, lactose, and xylose by recombinant Escherichia coli. *Applied and Environmental Microbiology, 55*(8), 1943–1948.
14. Orencio-Trejo, M., Utrilla, J., Fernández-Sandoval, M. T., Huerta-Beristain, G., Gosset, G., & Martinez, A. (2010). Engineering the Escherichia coli fermentative metabolism. *Biosystems Engineering II*, 71–107.

15. Yomano, L. P., York, S. W., & Ingram, L. O. (1998). Isolation and characterization of ethanol-tolerant mutants of Escherichia coli KO11 for fuel ethanol production. *Journal of Industrial Microbiology and Biotechnology, 20*(2), 132–138.
16. Qureshi, N., Dien, B., Nichols, N. N., Liu, S., Hughes, S., Iten, L. B., Saha, B. C., & Cotta, M. A. (2004, November). Continuous production of ethanol in high productivity bioreactors using genetically engineered Escherichia coli FBR5: membrane and fixed cell reactors. In *American Institute Of Chemical Engineers Annual Meeting.*
17. Ingram, L. O., Conway, T., & Alterthum, F. (1991). *U.S. Patent No. 5,000,000.* Washington, DC: U.S. Patent and Trademark Office.
18. Vinuselvi, P., & Lee, S. K. (2012). Engineered Escherichia coli capable of co-utilization of cellobiose and xylose. *Enzyme and Microbial Technology, 50*(1), 1–4.
19. Geddes, C. C., Mullinnix, M. T., Nieves, I. U., Peterson, J. J., Hoffman, R. W., York, S. W., Yomano, L. P., Miller, E. N., Shanmugam, K. T., & Ingram, L. O. (2011). Simplified process for ethanol production from sugarcane bagasse using hydrolysate-resistant Escherichia coli strain MM160. *Bioresource Technology, 102*(3), 2702–2711.
20. Nieves, I. U., Geddes, C. C., Mullinnix, M. T., Hoffman, R. W., Tong, Z., Castro, E., Shanmugam, K. T., & Ingram, L. O. (2011). Injection of air into the headspace improves fermentation of phosphoric acid pretreated sugarcane bagasse by Escherichia coli MM170. *Bioresource Technology, 102*(13), 6959–6965.
21. Zhang, M., Eddy, C., Deanda, K., Finkelstein, M., & Picataggio, S. (1995). Metabolic engineering of a pentose metabolism pathway in ethanologenic Zymomonas mobilis. *Science, 267*(5195), 240–243.
22. Parker, C., Barnell, W. O., Snoep, J. L., Ingram, L. O., & Conway, T. (1995). Characterization of the Zymomonas mobilis glucose facilitator gene product (glf) in recombinant Escherichia coli: Examination of transport mechanism, kinetics and the role of glucokinase in glucose transport. *Molecular Microbiology, 15*(5), 795–802.
23. Mohagheghi, A., Evans, K., Chou, Y. C., & Zhang, M. (2002). Cofermentation of glucose, xylose, and arabinose by genomic DNA-lntegrated xylose/arabinose fermenting strain of zymomonas mobilis ax101. In *Biotechnology for fuels and chemicals* (pp. 885–898). Humana Press, Totowa, NJ.
24. Lawford, H. G., & Rousseau, J. D. (2003). Cellulosic fuel ethanol. In *Biotechnology for fuels and chemicals* (pp. 457–469). Humana Press, Totowa, NJ.
25. Doran, J. B., & Ingram, L. O. (1993). Fermentation of crystalline cellulose to ethanol by Klebsiella oxytoca containing chromosomally integrated Zymomonas mobilis genes. *Biotechnology Progress, 9*(5), 533–538.
26. Jarboe, L. R., Grabar, T. B., Yomano, L. P., Shanmugan, K. T., & Ingram, L. O. (2007). Development of ethanologenic bacteria. *Biofuels*, 237–261.
27. Zhou, S., & Ingram, L. O. (2001). Simultaneous saccharification and fermentation of amorphous cellulose to ethanol by recombinant Klebsiella oxytoca SZ21 without supplemental cellulase. *Biotechnology Letters, 23*(18), 1455–1462.
28. Hamilton-Brehm, S. D., Mosher, J. J., Vishnivetskaya, T., Podar, M., Carroll, S., Allman, S., Phelps, T. J., Keller, M., & Elkins, J. G. (2010). Caldicellulosiruptor obsidiansis sp. nov., an anaerobic, extremely thermophilic, cellulolytic bacterium isolated from Obsidian Pool, Yellowstone National Park. *Applied and Environmental Microbiology, 76*(4), 1014–1020.
29. Lynd, L. R., Van Zyl, W. H., McBride, J. E., & Laser, M. (2005). Consolidated bioprocessing of cellulosic biomass: An update. *Current Opinion in Biotechnology, 16*(5), 577–583.
30. Shaw, A. J., Podkaminer, K. K., Desai, S. G., Bardsley, J. S., Rogers, S. R., Thorne, P. G., Hogsett, D. A., & Lynd, L. R. (2008). Metabolic engineering of a thermophilic bacterium to produce ethanol at high yield. *Proceedings of the National Academy of Sciences, 105*(37), 13769–13774.
31. Sai Ram, M., & Seenayya, G. (1989). Ethanol production byClostridium thermocellum SS8, a newly isolated thermophilic bacterium. *Biotechnology Letters, 11*(8), 589–592.
32. Warnick, T. A., Methe, B. A., & Leschine, S. B. (2002). Clostridium phytofermentans sp. nov., a cellulolytic mesophile from forest soil. *International Journal of Systematic and Evolutionary Microbiology, 52*(4), 1155–1160.

33. Kim, E. S., Lee, H. J., Bang, W. G., Choi, I. G., & Kim, K. H. (2009). Functional characterization of a bacterial expansin from Bacillus subtilis for enhanced enzymatic hydrolysis of cellulose. *Biotechnology and Bioengineering, 102*(5), 1342–1353.
34. Zhang, X. Z., Zhang, Z., Zhu, Z., Sathitsuksanoh, N., Yang, Y., & Zhang, Y. H. P. (2010). The noncellulosomal family 48 cellobiohydrolase from Clostridium phytofermentans ISDg: Heterologous expression, characterization, and processivity. *Applied Microbiology and Biotechnology, 86*(2), 525–533.
35. Knoshaug, E. P., & Zhang, M. (2009). Butanol tolerance in a selection of microorganisms. *Applied Biochemistry and Biotechnology, 153*(1), 13–20.
36. Weber, C., Farwick, A., Benisch, F., Brat, D., Dietz, H., Subtil, T., & Boles, E. (2010). Trends and challenges in the microbial production of lignocellulosic bioalcohol fuels. *Applied Microbiology and Biotechnology, 87*(4), 1303–1315.
37. Raman, B., McKeown, C. K., Rodriguez, M., Brown, S. D., & Mielenz, J. R. (2011). Transcriptomic analysis of Clostridium thermocellumATCC 27405 cellulose fermentation. *Bmc Microbiology, 11*(1), 1–15.
38. Oh, B. R., Seo, J. W., Heo, S. Y., Hong, W. K., Luo, L. H., Joe, M. H., Park, D. H., & Kim, C. H. (2011). Efficient production of ethanol from crude glycerol by a Klebsiella pneumoniae mutant strain. *Bioresource Technology, 102*(4), 3918–3922.
39. Xiao, H., Gu, Y., Ning, Y., Yang, Y., Mitchell, W. J., Jiang, W., & Yang, S. (2011). Confirmation and elimination of xylose metabolism bottlenecks in glucose phosphoenolpyruvate-dependent phosphotransferase system-deficient Clostridium acetobutylicum for simultaneous utilization of glucose, xylose, and arabinose. *Applied and Environmental Microbiology, 77*(22), 7886–7895.
40. Vasan, P. T., Piriya, P. S., Prabhu, D. I. G., & Vennison, S. J. (2011). Cellulosic ethanol production by Zymomonas mobilis harboring an endoglucanase gene from Enterobacter cloacae. *Bioresource Technology, 102*(3), 2585–2589.
41. Park, Y. C., Jun, S. Y., & Seo, J. H. (2012). Construction and characterization of recombinant Bacillus subtilis JY123 able to transport xylose efficiently. *Journal of Biotechnology, 161*(4), 402–406.
42. Kurosawa, K., Wewetzer, S. J., & Sinskey, A. J. (2014). *Triacylglycerol production from corn stover using a xylose-fermenting Rhodococcus opacus strain for lignocellulosic biofuels.*
43. Lynd, L. R., Weimer, P. J., Van Zyl, W. H., & Pretorius, I. S. (2002). Microbial cellulose utilization: Fundamentals and biotechnology. *Microbiology and Molecular Biology Reviews, 66*(3), 506–577.
44. Bhatia, S. K., Lee, B. R., Sathiyanarayanan, G., Song, H. S., Kim, J., Jeon, J. M., Kim, J. H., Park, S. H., Yu, J. H., Park, K., & Yang, Y. H. (2016). Medium engineering for enhanced production of undecylprodigiosin antibiotic in Streptomyces coelicolor using oil palm biomass hydrolysate as a carbon source. *Bioresource Technology, 217*, 141–149.
45. Olofsson, K., Palmqvist, B., & Lidén, G. (2010). Improving simultaneous saccharification and co-fermentation of pretreated wheat straw using both enzyme and substrate feeding. *Biotechnology for Biofuels, 3*(1), 1–9.
46. Lopes, M. L., Paulillo, S. C. D. L., Godoy, A., Cherubin, R. A., Lorenzi, M. S., Giometti, F. H. C., Bernardino, C. D., Amorim Neto, H. B. D., & Amorim, H. V. D. (2016). Ethanol production in Brazil: a bridge between science and industry. *Brazilian Journal of Microbiology, 47*, 64–76.
47. Van Duren, I., Voinov, A., Arodudu, O., & Firrisa, M. T. (2015). Where to produce rapeseed biodiesel and why? Mapping European rapeseed energy efficiency. *Renewable Energy, 74*, 49–59.
48. Bhatia, S. K., Bhatia, R. K., & Yang, Y. H. (2016). Biosynthesis of polyesters and polyamide building blocks using microbial fermentation and biotransformation. *Reviews in Environmental Science and Bio/Technology, 15*(4), 639–663.

Chapter 12
Ethanol Production from Xylose Through GM *Saccharomyces cerevisiae*

Abstract The second most important fermentable sugar found in lignocellulosic hydrolysates is xylose (pentose). Sugars such as C_6 and C_5 are used to produce a large amount of C_2H_5OH from lignocellulosic hydrolysates. The well-known yeast *Saccharomyces cerevisiae* is the desired microorganism for C_2H_5OH manufacturing. However, surprisingly, *S. cerevisiae* is not capable of using or fermenting xylose naturally. Over the last 15 years, it has aimed to increase yeast's potential to consume xylose and ferment it into ethanol. Xylose metabolism or uptake, redox imbalance, xylulokinase overexpression, and PPP have all been used to engineer yeasts capable of generating C_2H_5OH from pentose substrate. This study examines the research on genetically engineered strains of *S. cerevisiae* and xylose fermentation into C_2H_5OH.

Keywords Ethanol · Xylose · *Saccharomyces cerevisiae* · Metabolic engineering

12.1 Introduction

The last several years have significantly focused on the production of C_2H_5OH from crop raw material to be used in gasoline blends to decrease petroleum expenditure. The production of bioethanol from raw material feedstock for gasoline blends is currently focused in North America on the transformation of starch from grain. On the transformation of starch from grain. Grain has traditionally been used for feed. With the increased use of corn as biomass for ethanol production, grain availability will be insufficient, causing grain prices to rise unnecessarily. Farmers, on the other hand, benefit from rising grain prices. The amount of lignocellulosic biomass available for conversion is much greater than that of food crops, and it can be collected with far less disruption to the food economy [1]. The breakdown of lignocelluloses in biomass yields a mixture of monomeric sugars such as glucose (hexose) and xylose (pentose). Hexoses have the highest concentrations of these, while other sugars have lower concentrations. However, hemicellulosic sugars like xylose or mannose may dominate [2]. The use of microorganisms that efficiently transform both glucose and xylose found in lignocellulosic hydrolysates into the useful final output, increases the production of any biomass bioconversion process. However, various microbes have the potential to convert glucose into ethanol, and pentose

N. K. Aggarwal et al., *Bioethanol Production*, Green Chemistry and Sustainable Technology, https://doi.org/10.1007/978-3-031-05091-6_12

conversion does not achieve maximum efficiency. In the early 1980s, the discovery by Canadian and US researchers of pentose fermenting yeasts was a breakthrough that attracted worldwide attention in C_5 fermentation [3]. Current research shows that the yield of C_2H_5OH manufactured from pentose is considerably less than the hexose fermentation. Due to the absence of xylose transport and enzymes required for a xylose metabolism, yeast *S. cerevisiae* cannot potentially ferment xylose to ethanol [4]. *S. cerevisiae* shows high hexose fermentation rate and ethanol production, so most researchers focus on improving its xylose (pentose) fermenting rate. Ethanol production from LCB is a feasible form of energy. In *S. cerevisiae*, xylose is transformed into xylulose with the help of two enzymes that use distinct cofactors, which lead to a redox-imbalance or, as a result, inhibit xylose fermentation. Cloning of xylitol dehydrogenase and xylose reductase, all of which are connected to the same coenzyme, and cloning of a xylose isomerase, which transforms xylose to xylulose, are the two most common approaches to solving this issue. Despite that, yeast engineering follows that approach to fermenting xylose but still does it slowly [5]. As a result, to improve the specific utilization of xylose as well as the rate and production of ethanol, genetic modifications were required: (i) aldose reductase, which transforms xylose to xylitol, is degraded; (ii) required enzyme and xylose transporters are overexpressed [6]. Cell performance, C_2H_5OH output, and the potency of recombinant strains have all been improved using evolutionary engineering [6]. An evolutionary engineering approach is analogous to metabolic engineering in that it searches for useful phenotypes by applying one or more suitable pressures. Engineered xylose-fermenting yeast is unable to use xylose up to $C_6H_{12}O_6$ finish. One possible reason for this case is that $C_6H_{12}O_6$, via Mig1, a significant and important transcription factor for the catabolic repression phase, silences the expression of genes important for the breakdown of xylose into simpler ones. When there is a high concentration of glucose, Mig1 travels quickly from the cytoplasm into the nucleus or sticks to the promoters of glucose repressible genes. When cells do not have enough glucose, Mig1 returns to the cytoplasm, releasing $C_6H_{12}O_6$ repression [7]. Despite the fact that *S. cerevisiae* does not use xylose as a natural substrate, *Pichia stipitis* [8] & *Pachysolen tannophilus* [9] may ferment it. Pentose phosphate pathway [10] can release NADPH for reductive biosynthesis process within cells and the formation of 5-carbonsugars for the formation of the nucleotides. The pathway has two distinct steps. The oxidative steptrans convert the 6-C, D-glucose 6P, to the 5-C, D-ribulose 5P, with the addition of carbon dioxide or NADPH. D-ribulose 5P is converted to D-ribose 5P, D-xylulose 5P, D-sedoheptulose 7P, D-erythrose 4P, D-fructose 6P, or D-glyceraldehyde 3P in the non-oxidative stage. Via D-xylulose, D-xylose and L-arabinose join the Pentose Phosphate Pathway (Fig. 12.1) until entering PPP; xylose is directly isomerized to xylulose in bacteria with the aid of xylose isomerase (XI). Xylose is converted to xylitol by xylose reductase and then oxidized to xylulose by xylitol dehydrogenase (XDH) in yeasts that ferment xylose [11]. Xylulokinase (XK) phosphorylates xylulose to xylulose 5-phosphate, which is then metabolized through glycolysis and the pentose phosphate pathway (Fig. 12.2). Due to the lack of xylose reductase and xylitol dehydrogenase activity, yeast *S. cerevisiae* cannot metabolize xylose, but it can consume the isomeric form xylulose. Pentose metabolism in

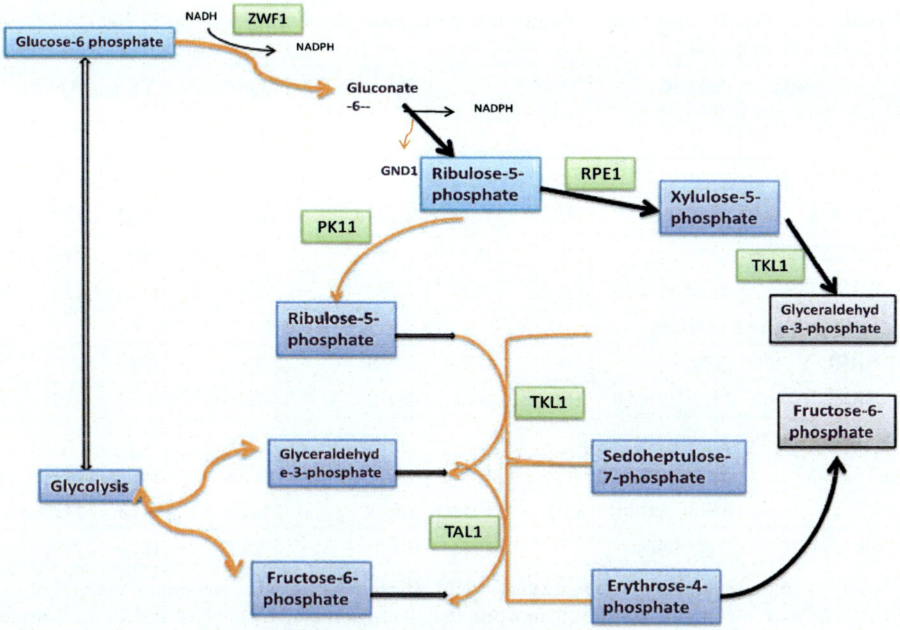

Fig. 12.1 Pentose phosphate pathway (*Source* Yang et al.) [112]

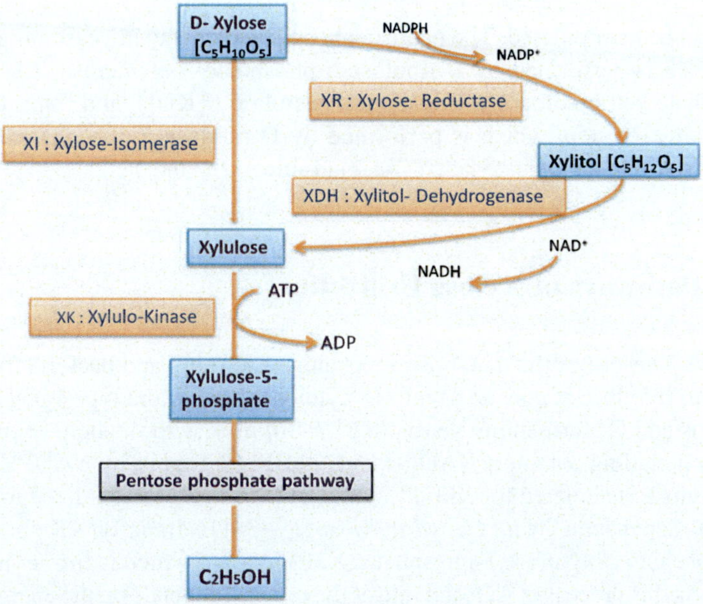

Fig. 12.2 Metabolism of xylose in bacteria (*Source Yang* et al.) [112]

Table 12.1 Performance in batch fermentation experiments of wild yeasts in the yield of C_2H_5OH or xylitol from xylose

Species/strain	Aeration	Xylose (g L^{-1})	Ye (g g^{-1})	q p, max (g g^{-1} L^{-1} h^{-1})	Q p, max (g L^{-1} h^{-1})	Yx (g g^{-1})	Reference
Candida shehatae							
Y12856	Aerobic	50	o.45	0.19	0.29	0.02	[12]
CBS 2779	Aerobic	50	0.37	0.48	0.96	0.08	[13]
CBS 2779	Semi-aerobic	50	0.34	0.28	0.74	0.13	[13]
Pachysolen tannophilus							
NRRL Y2460	Aerobic	100	0.25	0.058	0.13	0.24	[12]
NRRL Y2460	Microaerobic	150	0.21	0.059	0.086	0.57	[12]
Pichia stipites							
Y7124	Aerobic	100	0.42	0.23	0.38	0.01	[12]
Y7124	Microaerobic	150	0.38	0.073	0.23	0.13	[12]
CBS 7126	Anaerobic	50	0.38	0.18	0.22	0.09	[13]

Abbreviations: *Ye* ethanol yield, *Yx* xylitol yield; *Qp* max, maximum volumetric ethanol productivity; *qp* **max, maximum specific ethanol productivity**; *A* aerobic; *AN* anaerobic; *L* oxygen-limited; *Micro* microaerobic; *SA* semi-aerobic; *NS* not stated
Source (*Chu* et al.) [113]

yeasts can be manipulated. The pentose phosphate pathway in yeast has an oxidative step, which is performed by D-ribulose-5-phosphate 3-epimerase (RPE1), ribose-5-phosphate ketol-isomerase (RKI1), transketolase (TKL1), and transaldolase, and a non-oxidative step, which is performed by D-ribulose-5-phosphate 3-epimerase (RPE1), ribose-5-phosphate keto (TAL1) (Tables 12.1 and 12.2).

12.2 Pathways of Xylose Utilization

Figure 12.3 describes the metabolic mechanisms in fungi and bacteria for the intake of xylose. D-xylose is converted into D-xylulose through the type I phase in almost all fungus and C_5 fermenting yeasts (e.g., *P. stipitis*), with the help of two oxidoreductases including cofactors NADPH and NADP + . That is, by NADPH-dependent XR, D-xylose is depleted to xylitol [30–33] or then xylitol is oxidized to D-xylulose by NAD^+ dependent *Xylitol Dehydrogenase* [32–34] Eventually, XK phosphorylates D-xylulose into D-xylulose 5-phosphate (X5P) [35, 36] which is further metabolized by PPP. As this process is ATP, the rate of the process depends on the energy charge as well as the phosphorylation capacity of a cell, since this step uses ATP. *S. cerevisiae lacks the ability to consume xylose, despite the presence of genes encoding Xylose reductase (YHR104w, GRE3)* [37], *Xylitol Dehydrogenase (YLR070c, ScXYL2)* [38], *and Xylulokinase (XKS1)* [39, 40] *in its genome. In S. cerevisiae, overexpression of*

Table 12.2 Performance in batch fermentation process with recombinant *S. cerevisiae* strains in the yield of C_2H_5OH or xylitol from xylose

Strain	Description	Aeration	Xylose (g L^{-1})	Ye (g g^{-1})	Qp, e (g L^{-1} h^{-1})	Yx (g g^{-1})	Reference
pRD1	XYL1/XYL2	Anaerobic	21.7	0.07	0.07	0.46	[14]
H158	XYL1/XYL2	Anaerobic	80	0	0	0.53	[15]
H158-pXks	XYL1/XYL2/XKS1	Anaerobic	80	0.22	NS	0	[16]
H1693	XYL1-XYL2 (integrant)	Aerobic	50	0	NS	0.28	[17]
TMB3250	XYL1/XYL2/XKS1	Anaerobic	50	0.3	0.21^{c}	0.3	[18]
TMB3260	XYL1/XYL2/XKS1	Anaerobic	50	0.3	0.21^{c}	0.13	[19]
FPL-YS10	XYL1	Anaerobic	20	0	0	0.55	[20]
H2674	XYL1/XYL2/XKS1	Anaerobic	50	0.18	Not started	0.53	[21]
TMB3265	XYL2/XKS1	Anaerobic	10	0.19	0.026^{g}	0	[22]
TMB3050	XYLAT/XKS1/TAL/TKL RKI/RPE/ΔGRE3 xylose adaptation	Anaerobic	50	0.29	NS	0.23	[23]
RWB217	XYLAP/XK/TAL/TKL/RKI/RPE/ΔGRE3	Anaerobic	20	0.43	Not started	0.06	[24]
H2490-4	XYL1/XYL2/XKS1 chemostat isolate	Aerobic	30	0.1	Not started	0.21	[25]
TMB3270	XYL1(K270M one copy)/XYL2/XKS1	Anaerobic	50	0.36	0.32^{h}	0.17	[26]
TMB3057	XYL1/XYL2/XK/TAL/TKL/RKI/RPE/ΔGRE3	Anaerobic	50	0.33	0.13^{i}	0.22	[27, 28]

Abbreviations: *Ye* 1ethanol yield; *Qp,e* volumetric ethanol productivity; *Yx* xylitol yield; *C, D, and M* **are complex, defined minimal** [29] and minimal media, respectively; *Micro* micro aerobic; *L* oxygen-limited; *NS* not stated; *XYL1 Pichia stipitis* xylose reductase; *XYL2 P.stipitis* xylitol dehydrogenase; *XYL3 P. stipites* xylulokinase; *XKS1 S. cerevisiae* xylulokinase; *GRE3 S. cerevisiae* xylose reductase; *PGI* phosphoglucose isomerase; *zwf1* 6-phosphogluconate dehydrogenase; *GND1* 6-phosphogluconate dehydrogenase; *SOL3* 6-phosphogluconolactonase; *CTH* cytoplasmic transhydrogenase from *Azotobactervinelandii*; *XYLAP Piromyces* sp. E2 xylose isomerase; *XYLAT Thermus thermophilus* xylose isomerase; *TAL* trans aldolase; *TKL* transketolase; *RKI* ribose 5-phosphate ketol-isomerase; *RPE* ribulose 5-phosphate epimerase

C = Calculated using 70 h CT and 25 mL cultivation volume (CV); G = Calculated using 72 h CT and 25 mL CV; H = Calculated using 70 h CT and 20 mL CV; i = Calculated using I = 100 h CT and 23 mL CV.

Source (*Chu* et al.) [113]

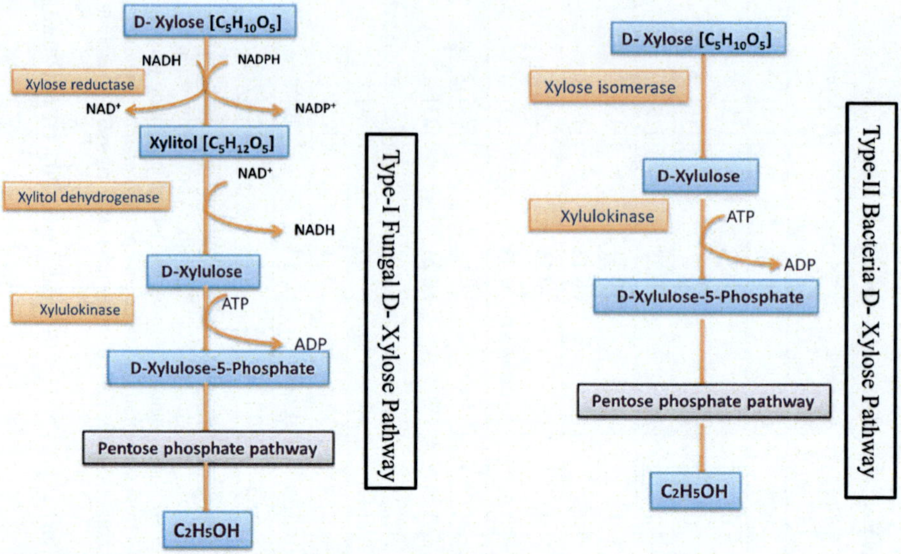

Fig. 12.3 D-xylose metabolic routes in fungi and bacteria

endogenous genes only slows development of xylose [41]. But in the case of bacteria such as *Escherichia coli* or *Streptomyces*, with the aid of xylose isomerase, D-xylose is directly converted into D-xylulose via the type-II route. Xylulokinase phosphorylated D-xylulose into D-xylulose 5-p, just like in fungi. Harbor XI activity has been recorded for both yeast [42], and some fungi [43, 44]

12.3 Xylose Reductase and Xylitol Dehydrogenase Expression

S. cerevisiae AN xylose fermentation was1stperform by heterologous expression of XYL1 [33] or XYL2 [45] genes that encode Xylose reductase or Xylitol dehydrogenase from *P. stipitis* (referred to as *P. stipitis* Xylose reductase and *P. stipitis* xylitol dehydrogenase) [14, 46]. In comparison to *S. cerevisiae*, *P. stipitis* converts xylose into C_2H_5OH at hypothetical production under AN conditions, it requires O_2 for potential C_5 use, and it has a low sugar utilization rate [8, 47]. Almost all other natural xylose-fermenting yeasts release xylitol. As a result, during the Xylose reductase and Xylitol dehydrogenase reactions, high concentrations of NADH accumulate due to a lack of NAD^+, resulting in an intracellular redox split [14, 48]. *P. stipites* contain both NADPH as well as NADH-specific Xylose reductase cofactors, allowing the yeast to emit low levels of xylitol during C_5 fermentation. While *P. tannophilus* is included in this cofactor dual XR, it generates greater xylitol than *P. stipitis* [49]. A significant quantity of xylitol is released in recombinant *S. cerevisiae* strains expressing *P. stipitis* Xylose reductase or *P. stipitis* xylitol dehydrogenase or xylitol decreases the C_2H_5OH amount [14, 46]. Recombinant *S. cerevisiae* xylose consumption rate is lower than glucose [50]. It has been shown that a higher level of Xylitol dehydrogenase compared to Xylose reductase decreases the amount of xylitol and increases the amount of C_2H_5OH [20, 27, 51, 52]. Current experimental studies show that high PsXR or PsXDH activity is also required for the production of a desirable xylose-fermenting recombinant *S. cerevisiae* strain [27, 53, 54].

12.4 XI Expression

The expression in *S. cerevisiae* of functional bacterial XI, encoded by the xylA gene, has made it challenging. *Actinoplanesmissouriensis* [55] and *Clostridium thermosulfurogenes* [56] have Heterologous expression of XI genes in *S. cerevisiae* produces inactive proteins in *S. cerevisiae* failed heterologous expression due to misfolding of protein, PTM, and disulfide bond formation. *Thermus thermophilus*1st functional XI was expressed in *S. cerevisiae* [57]. *Piromyces* [58] or Orpinomyces [59] XI genes are also expressed functionally in *S. cerevisiae* at a fast rate, but with very sluggish growth of C_5. In *S. cerevisiae*, some bacterial or fungal XI routes have also been discovered. A research reported effective expression of active bacterial XI from *Clostridium phytofermentans* shows decreased repression by xylitol in *S. cerevisiae* [60]. Using the XI route substitute of the Xylose reductase-Xylitol dehydrogenase route to reduce xylitol yield [57, 58, 61]. XI expression does not need redox cofactors or does not release an intracellular redox imbalance at the time of xylose fermentation. Direct comparison of the Xylose fermenting potential showed a low xylitol yield or high C_2H_5OH production in the XI-gene expressing recombinant *S. cerevisiae* strain compared to the Xylitol dehydrogenase-Xylose reductase express strain [28]. To tackle these problems, any other driving force is required while needed.

12.5 Factors Limiting Xylose Metabolism in *S. cerevisiae*

C_2H_5OH with a high rate and unique production should be produced by a model microbe that is visualized to be used in raw material change; have a large biomass range or high C_2H_5OH tolerance; or tolerance to lignocellulosic hydrolysates inhibitors [62]. *S. cerevisiae* is a desirable microorganism for commercial-scale ethanol production. It uses xylose as a carbon source ineffectively [4, 63]. Recombinant *S. cerevisiae* strains expressing the *P. stipitis XYL1* gene will effectively use xylose, while much of the xylose is transformed into xylitol [64–66]. In the first indication, this recombinant strain acts as a begging strain for the establishment of other mutant *S. cerevisiae* strains or enhances the rate of xylose consumption by metabolic engineering. *P. stipitis* genes XYL1 or XYL2 were inserted into *S. cerevisiae* TMB3001 to create a recombinant strain. Recognizing the rate-limiting reaction in pentose consumption by *S. cerevisiae* is critical for C_2H_5OH production. *S. cerevisiae that is unable to grow on xylose has been allotted to be incapable in the uptake of xylose* [14], *and in the first two phases of xylose metabolism, a redox imbalance* [67], *insufficient XK activity, and an imbalance have developed* Pentose phosphate pathway [68]. *S. cerevisiae* cannot grow on xylose due to repression of pentose sugar metabolism to proceed with the lower phase of glycolysis [69, 70].

12.5.1 Uptake of Xylose

Optimization of both uptakes of xylose or metabolism will be required in order to achieve industrial production of relevant ethanol under large extracellular xylose amounts. As a result, increasing xylose uptake in *S. cerevisiae* is as essential as optimizing the xylose metabolic process. In *S. cerevisiae*, xylose is transported by facilitated diffusion along with transporters from the Hxt family, but with a much lower potential than hexose. *S. cerevisiae* TMB3201 was engineered to use the identical plasmid as the TMB3001 strain, and has a breakdown of 18 HXT genes and is, therefore, not capable of developing two percent xylose [71, 72]. *S. cerevisiae also uses xylose as the sole carbon source for the expression of Hxt5 and Hxt7, as well as two Hxt proteins involved in xylose transport* [71, 73]. *Individual Hxt or linked proteins were studied during co-fermentation with pentose 4% & hexose 8% in a S. cerevisiae Hxt° mutant strain (RE700A)* [74]. *Profiles of mRNA copy or acquisition of intracellular xylose were studied in mutant strains with independent HXT1-5, 7 orgalactose transporter (GAL2) genes. According to the report, as the concentration of hexose declined and pentose utilization improved, high levels of HXT7 or raised levels of HXT5 or alpha glucoside transporter 1(AGT1) were observed* [74]. *HXT2 transcript levels were slowly triggered by hexose but quickly declined before the pentose use process because HXT1, HXT3, & HXT4 repress expression.* The significance of Hxt5 & Hxt7 for pentose uptake in *S. cerevisiae* is confirmed by this research work. Besides, the authors established the significance of the gene AGT1 for xylose uptake. Before participating in xylose uptake [71], the Gal2 gene had an inside cell pentose concentration that was less than Hxt5 or Hxt7 or its overexpression releases. The above study data is not surprising as the absence of xylose-specific uptakes or alternatives in *S. cerevisiae* absent depends on the Hxt family of proteins that have lower pentose help. In the slower pentose metabolizing strain TMB3001, overexpression of significant HXT7 or GAL2 xylose transporter did not enhance pentose consumption [71]. In addition, the reorganization of genes that have sugar-proton-symporters participate in pentose transport in *C.intermedia* (GXS1) or *S. cerevisiae*(AGT1) represents the capacity to develop specific goals for overexpression results.

12.5.2 Redox Imbalance from Xylose Reductase and Xylitol Dehydrogenase

Xylose enters into a yeast cell, and is changed over to the pentose phosphate pathway intermediate xylulose via a 2-phase isomerization performed by xylose reductase and xylitol dehydrogenase. The enzymes XR and XDH have been characterized or identified because of the key role that they play in the metabolism of xylose. Kuhn purified or identified the first XR from ATCC 26602 (from *S. cerevisiae*) [75]. The purified monomeric Xylose reductase was NADPH-dependent with a clear molecular weight of 37 kDa. *S. cerevisiae* xylose reductase was categorized as an aldose-ketose reductase, on the basis of amino acid composition or N-terminal chain details, but was immunologically unidentical to other yeast xylose reductases [75]. The recombinant gene product identifies the cloned *S. cerevisiae* XR, GRE3, or two different groups [76–78]. In each occurrence, the xylose reductase gene product was almost the similar molecular weight of thirty-seven-kDa or had close affinities for xylose (Km range of 14.8–27.9 mM) [75–77] as the native *S. cerevisiae* XR. The affinity of the *S. cerevisiae* XR for xylose was estimated positively with xylose reductase isolated from alternate yeasts like xylose reductase (isoformA) from *P. tannophilusCBS4044* (Km = 14 mM) [79]. When *S. cerevisiae* is grown on xylose mix biomass, this could relatively clarify the lower levels of internal growing SCXR activity notice [80]. Xylitol dehydrogenase is classified as a member of the alcohol dehydrogenase family (medium-chain) or shares structural properties with sorbitol dehydrogenases (medium-chain) [81]. Purification and identification of XDHs from different xylose-fermenting yeasts, such as *P. tannophilus* [82], *P. stipitis* [14, 83]. The *P. stipitis* NAD-dependent xylitol dehydrogenase has the same kinetic constants compared to the *S. cerevisiae* XDH [84]. *P. stipitis* NAD-dependent XDH is balanced against the overall concentrations of NAD and NADH by the catabolic reduction-charge [NADH/(NADH + NAD)] [83, 84]. The activation of NAD-dependent Xylitol dehydrogenase will decrease in cells with high concentrations of NADH, resulting in $C_5H_{12}O_5$ synthesis. Xylose reductase specificity for NADPH or Xylitol dehydrogenase results in a redox irregularity examined in normal pentose fermenting yeasts. To this end, the presence of O_2 xylose metabolism in yeast is regulated, which can influence yield or duration of ethanol production [4, 85]. The presence of oxygenation estimates the change in pentose C-flux between C_2H_5OH manufacture and substrate. During Xylose assimilation, production of ethanol is supported under anaerobic conditions, however, substrate formation is supported under aerobic conditions [85]. Aeration favors the inside cell redox balance between two cofactor (NADH/NAD or NADPH/NADP) systems. Due to lack of pyridine-nucleotide-transhydrogenase activity, these two cofactors have various roles in xylose yeast metabolism [48, 86].During the growth of xylose in the presence of oxygen, the rates of utilization or formation of both NADH/NAD or NADPH/NADP are equivalence, avoiding a decrease in nucleotide pools [87]. While e-acceptor O_2 is not present, xylose fermentation occurs, followed mainly by a NADH/NAD or NADPH/NADP cofactor pool redox deflcit. Redox-imbalance occurs separately in yeasts with xylose reductase or xylitol dehydrogenase reactions particularly to the NADPH or NAD cofactor [30]. During these oxidation or reduction reactions, the specific cofactor results in NADP or NADH accumulation of oxygen, NADPH can grow again through fructose-6-Phosphate in the Pentose phosphate pathway or the NADP-dependent enzymes, which decrease the NADP aggregation [86, 87]. While an insufficient amount of NAD is rearranged for the xylitol dehydrogenase reaction, an excess amount of NADPH relative to NAD is produced for the Xylose reductase or xylitol dehydrogenase catalyzed reactions, respectively. Due to deficiency of oxygen, the e- transport system is not able to re-oxidation NADH to NAD by respiration or net yield of NAD does not produce net yield by fermentation of ethanol [87, 88]. Regeneration of NAD for the xylitol dehydrogenase reaction can be efficient from the dihydroxyacetone to glycerol reduction [26]. Although the formation of glycerol is undesirable because it distracts carbon from ethanol manufacture, it is essential for the production of ethanol. This cofactor rearrangement imbalance occurs under situations of decreased respiration or contributes to the suppression of the activity of xylitol dehydrogenase, which significantly increases the formation of supply to xylitol and decreases the production of C_2H_5OH in yeast [30, 31, 82, 89]. PSXR and *P. stipitis* xylitol dehydrogenase (PSXDH) heterologous expression in *S. cerevisiae* permit growth of pentose with $C_5H_{12}O_5$ produced [14]. *P. stipitis xylose* reductase differs from *S. cerevisiae* xylose reductase in that it has two cofactors, one of which is NADPH or NADH. According to the current study, high Xylose reductase and Xylitol dehydrogenase activities are required to increase C_2H_5OH production while continuously decreasing xylitol yield from pentose [27]. Future efforts will be made to increase the rate of C_5 consumption in individual XR or XDH processes in order to improve ethanol production even further, Saleh et al. [90] developed genetically modified *S. cerevisiae* strains that possess both expressions of NADH-dependent PSXR and *P. stipitis* xylitol dehydrogenase. When genetically modified *S. cerevisiae* was introduced, strains containing mutant NADH-favor *P. stipites* xylose reductase were able to xylose ferment into C_2H_5OH, with a C_5 (1.5%) C_6 (0.5%) mixture producing the best strain of 0.43 g/g/1. Another *P. stipitis* xylitol dehydrogenase variant with a declining potential for NADPH close to NADH [91] was expressed in *S. cerevisiae* recombinant strain (strains TMB3270 and 3271) with *P. stipitis* xylitol dehydrogenase or endogenous xylulokinase [26]. Ethanol production of 0.40 g/g/1 was obtained for both strains on a one percent C_5–C_6 combination. These enzymes or genetically modified *S. cerevisiae* strains can be used in future research that requires additional changes to the initial pentose metabolism to increase ethanol production.

12.5.3 Xylulokinase

S. cerevisiae has the potential to very slowly consuming xylulose, an XR or XDH product, into C_2H_5OH [92]. *S. cerevisiae ATCC 24,860 fermented 5% xylulose to C_2H_5OH and xylulose was used much lower (0.035 g gl cells hl) than hexose (0.833 g gl cells hl)* [93]. *Use of xylulose is slow due to declining levels of inside* *cell xylulokinase response, which may partially restrict pentose consumption* [35, 94]. *S. cerevisiae XKS1 gene is located on right arm chromosome VII and encodes Xylulokinase (SCXK)* [36]. Ho et al. [95] were the first to demonstrate the potential of recombinant *S. cerevisiae* strains for efficient C_5 fermentation and pentose-hexose co-fermentation. This strain was specified because it was the 1st to overexpress *S. cerevisiae* Xylulokinase, from a large copy of shuttle plasmid (yeast–*E. coli*), along with the PSXR and PSXD. Toivari et al. [17] also published successful xylose ferment C2H5OH by overexpression of XKS1 in an unidentified *S. Cerevisiae* PSXR/PSXD strain. Integrant strain overexpression of SCXK continuously increases the specific rate of xylose consumption. Furthermore, using MSA medium and pentose as the sole carbon source, integrant strain xylose converts to ethanol in the presence or absence of oxygen.

12.5.4 Non-oxidant Pentose Phosphate Enzymes Transketolase or Transaldolase

Fermentation of xylulose by S. cerevisiae is slowed due to inefficient use of Pentose phosphate pathway metabolites [14, 68, 96]. As a rate-control for fermentation of xylose or xylulose, 2 enzymes of the non-oxidative pentose phosphate pathway, TKL (transketolase) or TAL, were involved. The aggregation of the Pentose phosphate pathway intermediate S7P through the fermentation of xylulose is proof to support this [97]. Increased amounts of gene *transketolase or transaldolase* expression in *S. cerevisiae* grown in the presence of sugars like C_5/C_6 [41]. So TKL converts Xylose-5-P or Ribose-5-P. Sedoheplose-7-P or Glyceraldehyde-3-P; TAL transforms Sedoheptulose-7-P or Glyceraldehyde-3-P into Fructose-6-P or Erytherose-4-P (Fig. 12.1) [68]. The intermediates, Glyceraldehyde-3-Phosphate or F6P, are shared by the pentose phosphate pathway or glycolysis, and any decline can lead to incomplete xylulose fermentation in *S. cerevisiae.* While improving the activity TAL, anaerobic growth on xylose did not result. Karhumaa et al. [23] developed a genetically modified *S. cerevisiae* (TMB3057) with the same genotype as TMB3026, but with PSXR/PSXD on a GRE3 deletion plasmid. Strain TMB3057 consistently improved its rate of growth (0.16 h1) on 5percent xylose, confirming the need to improve the expression of transaldolase or other non-oxidative Pentose phosphate pathway enzymes for xylose consumption, as well as the activity of xylose reductase, xylitol dehydrogenase, and Xylulokinase for both xylose consumption and fermentation [27].

12.6 Host Strain Selection for Xylose Metabolized Fermentation

Determining the host strain for xylose *S. cerevisiae* consumption desirable for pentose fermentation into C_2H_5OH is most important for process optimization as the potential for xylulose fermentation or the ability to tolerate the inhibit or in lignocellulosic hydrolysates varies between desirable output strains [98, 99]. Until now, many different commercial or laboratory *Saccharomyces* strains have been metabolically modified to increase xylose consumption and compare their fermentation efficiency to xylose or hydrolysates [100–103]. For molecular genetics research, laboratory strains are desirable. However, commercial strains are normally chosen based on optimal results on an industrial scale. Laboratory strains tolerate hydrolysates poorly, whereas commercial *S. cerevisiae* strains tolerate them better [101, 104, 105]. *S. cerevisiae* strains have variant xylulose fermentation potential [93, 103, 106, 108], suggesting that there are native variations in their potential to ferment C_5 sugars. The variation in xylulose fermentation potential is mostly due to the difference in the movement of the Pentose phosphate pathway connecting the xylose-to-xylulose pathway to the glycolysis pathway [107]. A flocculent commercial strain of *S. cerevisiae* IR-2 was discovered in the current research scenario, with the highest potential for xylulose fermentation among mostly various commercial diploid strains for genetically modifying xylose fermentation [108]. The MA-R4 strain, genetically engineered from IR-2 with the help of chromosomal integration to illustrate the *P. stipitis* xylose reductase/*P. stipitis* xylitol dehydrogenase genes as well as the ScXK gene show desirable xylose-fermenting potential in comparison to many other recombinant commercial strains [103]. One more valuable engineering strategy aimed at the absence of oxygen fermentation of xylose can be used to estimate the potential for fermenting xylulose.

12.7 Modification of Xylose-Efficient *S. cerevisiae* Strains

A desirable method of *S. cerevisiae* for xylose fermentation was performed through selective pressure. Three different adaptation strategies were used, the first two using genetically modified *S. cerevisiae* strains, the third using a natural strain. In the presence of oxygen, the 1st adaptive strategy involves increasing the recombinant strain and moderately transforming the strain into the AN environment [109]. Two different populations, with distinct phenotypes, were developed by this strategy. The small population (class I) shows improved transport rates of xylose or could ferment sugar, such as ethanol mixtures of pentose-hexose (xylose/glucose). However, the large class II can grow AN on xylose [109]. The 2nd adaptive strategy includes a genetically modified strain only on pentose, then diverts to a combination of pentose-hexose and diverts back to pentose [25]. This strategy resulted in the development of an oxygen-free *S. cerevisiae* strain able to ferment 4.5% xylose to C_2H_5OH yield of 0.14 g/g/1 [25]. Metabolite concentrations inside the cell from chemostat isolates were measured in the presence or absence of oxygen. The results show that xylose metabolism was shown by improving glyoxylate cycle metabolite levels, glyoxylate (GLO), improving ICL (isocitrate lyase) activity, as well as enhancing tricarboxylic acid cycle malate levels. The tricarboxylic acid cycle may be limited to the fermentation of xylose due to malate aggregation. Jin et al. [110] also examined that under the control of O_2, recombinant *S. cerevisiae* cells expressed to have tricarboxylic acid cycle catalyst at raised levels, along with regulatory proteins or respiratory catalyst. The findings show that *S. cerevisiae* does not use xylose as $C_6H_{12}O_6$, and thus is not capable of silent respiration or other related processes. Large-scale xylose downstream reactions of xylulokinase, xylitol dehydrogenase, and xylose reductase may be needed for pentose fermentation in *S. cerevisiae* to achieve the C_2H_5OH production required for industrial ethanol processes. The third beneficial strategy is based on the current observation that *S. cerevisiae* mutants, which can grow slowly on pentose, continually appear under selection pressure. After weeks of incubation, Attfield and Bell [111] get miniature *S. cerevisiae* colonies on xylose medium. Strain MBG-2303 was able to grow on five percent xylose presence of oxygen with little amounts of xylitol or glycerol and a maximum of 0.58 g L^{-1} ethanol manufactured.

12.8 Summary

The main product of the breakdown of hemicelluloses is D-xylose and significant research attempts have been concentrated on the improvement of recombinant *S. cerevisiae* that has the potential for xylose fermentation. Significant advances have been made in the production of xylose ethanol by metabolic engineering. There are also unexplained measures in xylose metabolism through metabolically engineered *S. cerevisiae*, like low C_2H_5OH production, high byproducts, and the condition of these recombinant strains. Multiple tasks remain in xylose metabolic engineering. In the laboratory, a recombinant strain of *S. cerevisiae* was created that has a low genetic background, growth, and physiological characteristics when compared to commercial yeast strains. More effort is required for the reconstruction of the natural type of yeast in order for strains to be commercialized. Improved stability of heterogeneous gene expression in yeast can be expected to result in increased ethanol yield.

References

1. Stephanopoulos, G. (2007). Challenges in engineering microbes for biofuels production. *Science, 315*(5813), 801–804.
2. Helle, S. S., Murray, A., Lam, J., Cameron, D. R., & Duff, S. J. (2004). Xylose fermentation by genetically modified *Saccharomyces cerevisiae* 259ST in spent sulfite liquor. *Bioresource Technology, 92*(2), 163–171.

3. Schneider, H., Wang, P. Y., Chan, Y. K., & Maleszka, R. (1981). Conversion of D-xylose into ethanol by the yeast *Pachysolen tannophilus*. *Biotechnology Letters, 3*(2), 89–92.
4. Batt, C. A., Caryallo, S., Easson, D. D., Jr., Akedo, M., & Sinskey, A. J. (1986). Direct evidence for a xylose metabolic pathway in *Saccharomyces cerevisiae*. *Biotechnology and Bioengineering, 28*(4), 549–553.
5. Kim, S. R., Ha, S. J., Wei, N., Oh, E. J., & Jin, Y. S. (2012). Simultaneous co-fermentation of mixed sugars: A promising strategy for producing cellulosic ethanol. *Trends in Biotechnology, 30*(5), 274–282.
6. Cai, Z., Zhang, B., & Li, Y. (2012). Engineering *Saccharomyces cerevisiae* for efficient anaerobic xylose fermentation: Reflections and perspectives. *Biotechnology Journal, 7*(1), 34–46.
7. Rolland, F., Winderickx, J., & Thevelein, J. M. (2002). Glucose-sensing and-signalling mechanisms in yeast. *FEMS Yeast Research, 2*(2), 183–201.
8. Jeffries, T. W., Grigoriev, I. V., Grimwood, J., Laplaza, J. M., Aerts, A., Salamov, A., Schmutz, J., Lindquist, E., Dehal, P., Shapiro, H., & Jin, Y. S. (2007). Genome sequence of the lignocellulose-bioconverting and xylose-fermenting yeast *Pichia stipitis*. *Nature Biotechnology, 25*(3), 319–326.
9. Slininger, P. J., Bothast, R. J., Van Cauwenberge, J. E., & Kurtzman, C. P. (1982). Conversion of D-xylose to ethanol by the yeast *Pachysolen tannophilus*. *Biotechnology and Bioengineering, 24*(2), 371–384.
10. Jeffries, T. W. (2006). Engineering yeasts for xylose metabolism. *Current Opinion in Biotechnology, 17*(3), 320–326.
11. Traff, K. L., Otero Cordero, R. R., Van Zyl, W. H., & Hahn-Hagerdal, B. (2001). Deletion of the GRE3 aldose reductase gene and its influence on xylose metabolism in recombinant strains of *Saccharomyces cerevisiae* expressing the xylA and XKS1 genes. *Applied and Environmental Microbiology, 67*(12), 5668–5674.
12. Slininger, P. J., Bothast, R. J., Okos, M. R., & Ladisch, M. R. (1985). Comparative evaluation of ethanol production by xylose-fermenting yeasts presented high xylose concentrations. *Biotechnology Letters, 7*(6), 431–436.
13. Du Preez, J. C., Bosch, M., & Prior, B. A. (1986). The fermentation of hexose and pentose sugars by *Candida shehatae* and *Pichia stipitis*. *Applied Microbiology and Biotechnology, 23*(3), 228–233.
14. Kötter, P., & Ciriacy, M. (1993). Xylose fermentation by *Saccharomyces cerevisiae*. *Applied Microbiology and Biotechnology, 38*(6), 776–783.
15. Van Zyl, W. H., Eliasson, A., Hobley, T., & Hahn-Hägerdal, B. (1999). Xylose utilisation by recombinant strains of *Saccharomyces cerevisiae* on different carbon sources. *Applied Microbiology and Biotechnology, 52*(6), 829–833.
16. Johansson, B., Christensson, C., Hobley, T., & Hahn-Hägerdal, B. (2001). Xylulokinase overexpression in two strains of *Saccharomyces cerevisiae* also expressing xylose reductase and xylitol dehydrogenase and its effect on fermentation of xylose and lignocellulosic hydrolysate. *Applied and Environmental Microbiology, 67*(9), 4249–4255.
17. Toivari, M. H., Aristidou, A., Ruohonen, L., & Penttilä, M. (2001). Conversion of xylose to ethanol by recombinant *Saccharomyces cerevisiae*: Importance of xylulokinase (XKS1) and oxygen availability. *Metabolic Engineering, 3*(3), 236–249.
18. Jeppsson, M., Johansson, B., Hahn-Hägerdal, B., & Gorwa-Grauslund, M. F. (2002). Reduced oxidative pentose phosphate pathway flux in recombinant xylose-utilizing *Saccharomyces cerevisiae* strains improves the ethanol yield from xylose. *Applied and Environmental Microbiology, 68*(4), 1604–1609.
19. Jeppsson, M., Träff, K., Johansson, B., Hahn-Hägerdal, B., & Gorwa-Grauslund, M. F. (2003). Effect of enhanced xylose reductase activity on xylose consumption and product distribution in xylose-fermenting recombinant *Saccharomyces cerevisiae*. *FEMS Yeast Research, 3*(2), 167–175.
20. Jin, Y. S., & Jeffries, T. W. (2003). Changing flux of xylose metabolites by altering expression of xylose reductase and xylitol dehydrogenase in recombinant Saccharomyces cerevisiae. In *Biotechnology for Fuels and Chemicals* (pp. 277–285). Humana Press.

21. Verho, R., Londesborough, J., Penttilä, M., & Richard, P. (2003). Engineering redox cofactor regeneration for improved pentose fermentation in *Saccharomyces cerevisiae. Applied and Environmental Microbiology*, *69*(10), 5892–5897
22. Träff-Bjerre, K. L., Jeppsson, M., Hahn-Hägerdal, B., & Gorwa-Grauslund, M. F. (2004). Endogenous NADPH-dependent aldose reductase activity influences product formation during xylose consumption in recombinant *Saccharomyces cerevisiae. Yeast, 21*(2), 141–150.
23. Karhumaa, K., Hahn-Hägerdal, B., & Gorwa-Grauslund, M. F. (2005). Investigation of limiting metabolic steps in the utilization of xylose by recombinant *Saccharomyces cerevisiae* using metabolic engineering. *Yeast, 22*(5), 359–368.
24. Kuyper, M., Hartog, M. M., Toirkens, M. J., Almering, M. J., Winkler, A. A., van Dijken, J. P., & Pronk, J. T. (2005). Metabolic engineering of a xylose-isomerase-expressing *Saccharomyces cerevisiae* strain for rapid anaerobic xylose fermentation. *FEMS Yeast Research, 5*(4–5), 399–409.
25. Pitkänen, J. P., Rintala, E., Aristidou, A., Ruohonen, L., & Penttilä, M. (2005). Xylose chemostat isolates of *Saccharomyces cerevisiae* show altered metabolite and enzyme levels compared with xylose, glucose, and ethanol metabolism of the original strain. *Applied Microbiology and Biotechnology, 67*(6), 827–837.
26. Jeppsson, M., Bengtsson, O., Franke, K., Lee, H., Hahn-Hägerdal, B., & Gorwa-Grauslund, M. F. (2006). The expression of a *Pichia stipitis* xylose reductase mutant with higher KM for NADPH increases ethanol production from xylose in recombinant *Saccharomyces cerevisiae. Biotechnology and Bioengineering, 93*(4), 665–673.
27. Karhumaa, K., Fromanger, R., Hahn-Hägerdal, B., & Gorwa-Grauslund, M. F. (2007). High activity of xylose reductase and xylitol dehydrogenase improves xylose fermentation by recombinant *Saccharomyces cerevisiae. Applied Microbiology and Biotechnology, 73*(5), 1039–1046.
28. Karhumaa, K., Sanchez, R. G., Hahn-Hägerdal, B., & Gorwa-Grauslund, M. F. (2007). Comparison of the xylose reductase-xylitol dehydrogenase and the xylose isomerase pathways for xylose fermentation by recombinant *Saccharomyces cerevisiae. Microbial Cell Factories, 6*(1), 1–10.
29. Verduyn, C., Postma, E., Scheffers, W. A., & Van Dijken, J. P. (1992). Effect of benzoic acid on metabolic fluxes in yeasts: A continuous-culture study on the regulation of respiration and alcoholic fermentation. *Yeast, 8*(7), 501–517.
30. Bruinenberg, P. M., de Bot, P. H., van Dijken, J. P., & Scheffers, W. A. (1984). NADH-linked aldose reductase: The key to anaerobic alcoholic fermentation of xylose by yeasts. *Applied Microbiology and Biotechnology, 19*(4), 256–260.
31. Verduyn, C., Van Kleef, R., Frank, J., Schreuder, H., Van Dijken, J. P., & Scheffers, W. A. (1985). Properties of the NAD (P) H-dependent xylose reductase from the xylose-fermenting yeast *Pichia stipitis. Biochemical Journal, 226*(3), 669–677.
32. Bolen, P. L., Roth, K. A., & Freer, S. N. (1986). Affinity purifications of aldose reductase and xylitol dehydrogenase from the xylose-fermenting yeast *Pachysolen tannophilus. Applied and Environmental Microbiology, 52*(4), 660–664.
33. Rizzi, M., Erlemann, P., Bui-Thanh, N. A., & Dellweg, H. (1988). Xylose fermentation by yeasts. *Applied Microbiology and Biotechnology, 29*(2), 148–154.
34. Yang, V. W., & Jeffries, T. W. (1990). Purification and properties of xylitol dehydrogenase from the xylose-fermenting yeast *Candida shehatae. Applied Biochemistry and Biotechnology, 26*(2), 197–206.
35. Deng, X. X., & Ho, N. W. (1990). Xylulokinase activity in various yeasts including *Saccharomyces cerevisiae* containing the cloned xylulokinase gene. *Applied Biochemistry and Biotechnology, 24*(1), 193–199.
36. Rodriguez-Peña, J. M., Cid, V. J., Arroyo, J., & Nombela, C. (1998). The YGR194c (XKS1) gene encodes the xylulokinase from the budding yeast *Saccharomyces cerevisiae. FEMS Microbiology Letters, 162*(1), 155–160.
37. Träff, K. L., Jönsson, L. J., & Hahn-Hägerdal, B. (2002). Putative xylose and arabinose reductases in *Saccharomyces cerevisiae. Yeast, 19*(14), 1233–1241.

38. Richard, P., Toivari, M. H., & Penttilä, M. (1999). Evidence that the gene YLR070c of *Saccharomyces cerevisiae* encodes a xylitol dehydrogenase. *FEBS Letters, 457*(1), 135–138.
39. Yang, V. W., & Jeffries, T. W. (1997). Regulation of phosphotransferases in glucose-and xylose-fermenting yeasts. In *Biotechnology for fuels and chemicals* (pp. 97–108). Humana Press.
40. Richard, P., Toivari, M. H., & Penttilä, M. (2000). The role of xylulokinase in *Saccharomyces cerevisiae* xylulose catabolism. *FEMS Microbiology Letters, 190*(1), 39–43.
41. Toivari, M. H., Salusjärvi, L., Ruohonen, L., & Penttilä, M. (2004). Endogenous xylose pathway in *Saccharomyces cerevisiae. Applied and Environmental Microbiology, 70*(6), 3681–3686.
42. Vongsuvanlert, V., & Tani, Y. (1988). Purification and characterization of xylose isomerase of a methanol yeast, *Candida boidinii*, which is involved in sorbitol production from glucose. *Agricultural and Biological Chemistry, 52*(7), 1817–1824.
43. Banerjee, S., Archana, A., & Satyanarayana, T. (1994). Xylose metabolism in a thermophilic mouldMalbranchea pulchella var. sulfurea TMD-8. *Current Microbiology, 29*(6), 349–352.
44. Harhangi, H. R., Akhmanova, A. S., Emmens, R., van der Drift, C., de Laat, W. T., van Dijken, J. P., Jetten, M. S., Pronk, J. T., & Op den Camp, H. J. (2003). Xylose metabolism in the anaerobic fungus *Piromyces* sp. strain E2 follows the bacterial pathway. *Archives of Microbiology, 180*(2), 134–141.
45. Rizzi, M., Harwart, K., Erlemann, P., Bui-Thanh, N. A., & Dellweg, H. (1989). Purification and properties of the NAD+ -xylitol-dehydrogenase from the yeast *Pichia stipitis. Journal of Fermentation and Bioengineering, 67*, 20–24.
46. Tantirungkij, M., Nakashima, N., Seki, T., & Yoshida, T. (1993). Construction of xylose-assimilating *Saccharomyces cerevisiae. Journal of Fermentation and Bioengineering, 75*(2), 83–88.
47. Agbogbo, F. K., & Coward-Kelly, G. (2008). Cellulosic ethanol production using the naturally occurring xylose-fermenting yeast, *Pichia stipitis. Biotechnology Letters, 30*(9), 1515–1524.
48. Bruinenberg, P. M., de Bot, P. H., van Dijken, J. P., & Scheffers, W. A. (1983). The role of redox balances in the anaerobic fermentation of xylose by yeasts. *European Journal of Applied Microbiology and Biotechnology, 18*(5), 287–292.
49. Debus, D., Methner, H., Schulze, D., & Dellweg, H. (1983). Fermentation of xylose with the yeast *Pachysolen tannophilus. European Journal of Applied Microbiology and Biotechnology, 17*(5), 287–291.
50. Hahn-Hägerdal, B., Karhumaa, K., Larsson, C. U., Gorwa-Grauslund, M., Görgens, J., & Van Zyl, W. H. (2005). Role of cultivation media in the development of yeast strains for large scale industrial use. *Microbial Cell Factories, 4*(1), 1–16.
51. Walfridsson, M., Anderlund, M., Bao, X., & Hahn-Hägerdal, B. (1997). Expression of different levels of enzymes from the *Pichia stipitis* XYL1 and XYL2 genes in *Saccharomyces cerevisiae* and its effects on product formation during xylose utilisation. *Applied Microbiology and Biotechnology, 48*(2), 218–224.
52. Eliasson, A., Hofmeyr, J. H. S., Pedler, S., & Hahn-Hägerdal, B. (2001). The xylose reductase/xylitol dehydrogenase/xylulokinase ratio affects product formation in recombinant xylose-utilising *Saccharomyces cerevisiae. Enzyme and Microbial Technology, 29*(4–5), 288–297.
53. Johansson, B., & Hahn-Hägerdal, B. (2002). Overproduction of pentose phosphate pathway enzymes using a new CRE–loxP expression vector for repeated genomic integration in *Saccharomyces cerevisiae. Yeast, 19*(3), 225–231.
54. Matsushika, A., & Sawayama, S. (2008). Efficient bioethanol production from xylose by recombinant *Saccharomyces cerevisiae* requires high activity of xylose reductase and moderate xylulokinase activity. *Journal of Bioscience and Bioengineering, 106*(3), 306–309.
55. Amore, R., Wilhelm, M., & Hollenberg, C. P. (1989). The fermentation of xylose—an analysis of the expression of Bacillus and Actinoplanes xylose isomerase genes in yeast. *Applied Microbiology and Biotechnology, 30*(4), 351–357.

56. Moes, C. J., Pretorius, I. S., & Van Zyl, W. H. (1996). Cloning and expression of the *Clostridium thermosulfurogenes* D-xylose isomerase gene (xylA) in *Saccharomyces cerevisiae. Biotechnology Letters, 18*(3), 269–274.
57. Walfridsson, M., Bao, X., Anderlund, M., Lilius, G., Bülow, L., & Hahn-Hägerdal, B. (1996). Ethanolic fermentation of xylose with *Saccharomyces cerevisiae* harboring the *Thermus thermophilus* xylA gene, which expresses an active xylose (glucose) isomerase. *Applied and Environmental Microbiology, 62*(12), 4648–4651.
58. Kuyper, M., Harhangi, H. R., Stave, A. K., Winkler, A. A., Jetten, M. S., de Laat, W. T., den Ridder, J. J., Op den Camp, H. J., van Dijken, J. P., & Pronk, J. T. (2003). High-level functional expression of a fungal xylose isomerase: The key to efficient ethanolic fermentation of xylose by *Saccharomyces cerevisiae*? *FEMS Yeast Research, 4*(1), 69–78.
59. Madhavan, A., Tamalampudi, S., Ushida, K., Kanai, D., Katahira, S., Srivastava, A., Fukuda, H., Bisaria, V. S., & Kondo, A. (2009). Xylose isomerase from polycentric fungus Orpinomyces: Gene sequencing, cloning, and expression in *Saccharomyces cerevisiae* for bioconversion of xylose to ethanol. *Applied Microbiology and Biotechnology, 82*(6), 1067–1078.
60. Brat, D., Boles, E., & Wiedemann, B. (2009). Functional expression of a bacterial xylose isomerase in *Saccharomyces cerevisiae. Applied and Environmental Microbiology, 75*(8), 2304–2311.
61. Lönn, A., Träff-Bjerre, K. L., Otero, R. C., Van Zyl, W. H., & Hahn-Hägerdal, B. (2003). Xylose isomerase activity influences xylose fermentation with recombinant *Saccharomyces cerevisiae* strains expressing mutated xylA from *Thermus thermophilus. Enzyme and Microbial Technology, 32*(5), 567–573.
62. Aristidou, A., & Penttilä, M. (2000). Metabolic engineering applications to renewable resource utilization. *Current Opinion in Biotechnology, 11*(2), 187–198.
63. Barnett, J. A. (1976). The utilization of sugars by yeasts. *Advances in Carbohydrate Chemistry and Biochemistry, 32*, 125–234.
64. Amore, R., Kötter, P., Küster, C., Ciriacy, M., & Hollenberg, C. P. (1991). Cloning and expression in *Saccharomyces cerevisiae* of the NAD (P) H-dependent xylose reductase-encoding gene (XYL1) from the xylose-assimilating yeast *Pichia stipitis. Gene, 109*(1), 89–97.
65. Hallborn, J., Walfridsson, M., Airaksinen, U., Ojamo, H., Hahn-Hägerdal, B., Penttilä, M., & Keränen, S. (1991). Xylitol production by recombinant *Saccharomyces cerevisiae. Bio/Technology, 9*(11), 1090–1095.
66. Takuma, S., Nakashima, N., Tantirungkij, M., Kinoshita, S., Okada, H., Sew, T., & Yoshida, T. (1991). Isolation of xylose reductase gene ofPichia stipitis and its expression in *Saccharomyces cerevisiae. Applied Biochemistry and Biotechnology, 28*(1), 327–340.
67. Bruinenberg, P. M. (1986). The NADP (H) redox couple in yeast metabolism. *Antonie van Leeuwenhoek, 52*(5), 411–429.
68. Walfridsson, M., Hallborn, J., Penttilä, M., Keränen, S., & Hahn-Hägerdal, B. (1995). Xylose-metabolizing *Saccharomyces cerevisiae* strains overexpressing the TKL1 and TAL1 genes encoding the pentose phosphate pathway enzymes transketolase and transaldolase. *Applied and Environmental Microbiology, 61*(12), 4184–4190
69. Boles, E., Lehnert, W., & Zimmermann, F. K. (1993). The role of the NAD-dependent glutamate dehydrogenase in restoring growth on glucose of a *Saccharomyces cerevisiae* phosphoglucose isomerase mutant. *European Journal of Biochemistry, 217*(1), 469–477.
70. Müller, S., Boles, E., May, M., & Zimmermann, F. K. (1995). Different internal metabolites trigger the induction of glycolytic gene expression in *Saccharomyces cerevisiae. Journal of Bacteriology, 177*(15), 4517–4519.
71. Wieczorke, R., Krampe, S., Weierstall, T., Freidel, K., Hollenberg, C. P., & Boles, E. (1999). Concurrent knock-out of at least 20 transporter genes is required to block uptake of hexoses in *Saccharomyces cerevisiae. FEBS Letters, 464*(3), 123–128.
72. Schepper, M., Diderich, J. A., van Hoek, W. P. M., Luttik, M. A. H., van Dijken, J. P., Pronk, J. T., Klaassen, P., De Mattos, J. T., van Dam, K., & Kruckeberg, A. L. (1998). Glucose uptake kinetics and transcription of HXT genes in chemostat cultures of *Saccharomyces cerevisiae.*

In *EC Framework IV Symposium Yeast as a Cell Factory* (pp. 37–38). Delft University of Technology.
73. Sedlak, M., & Ho, N. W. (2004). Characterization of the effectiveness of hexose transporters for transporting xylose during glucose and xylose co-fermentation by a recombinant Saccharomyces yeast. *Yeast, 21*(8), 671–684.
74. Hamacher, T., Becker, J., Gárdonyi, M., Hahn-Hägerdal, B., & Boles, E. (2002). Characterization of the xylose-transporting properties of yeast hexose transporters and their influence on xylose utilization. *Microbiology, 148*(9), 2783–2788.
75. Kuhn, A., van Zyl, C., van Tonder, A. N. D. R. É., & Prior, B. A. (1995). Purification and partial characterization of an aldo-keto reductase from *Saccharomyces cerevisiae. Applied and Environmental Microbiology, 61*(4), 1580–1585.
76. Ford, G., & Ellis, E. M. (2001). Three aldo–keto reductases of the yeast *Saccharomyces cerevisiae. Chemico-Biological Interactions, 130*, 685–698.
77. Ye Jeong, E., Sopher, C., Seon Kim, I., & Lee, H. (2001). Mutational study of the role of tyrosine-49 in the *Saccharomyces cerevisiae* xylose reductase. *Yeast, 18*(11), 1081–1089.
78. Jeong, E. Y., Kim, I. S., & Lee, H. (2002). Identification of lysine-78 as an essential residue in the *Saccharomyces cerevisiae* xylose reductase. *FEMS Microbiology Letters, 209*(2), 223–228.
79. Ditzelmüller, G., Kubicek, C. P., Wöhrer, W., & Röhr, M. (1984). Xylose metabolism in *Pachysolen tannophilus*: Purification and properties of xylose reductase. *Canadian Journal of Microbiology, 30*(11), 1330–1336.
80. Van Zyl, C., Prior, B. A., Kilian, S. G., & Kock, J. L. (1989). D-xylose utilization by *Saccharomyces cerevisiae. Microbiology, 135*(11), 2791–2798.
81. Lunzer, R., Mamnun, Y., Haltrich, D., Kulbe, K. D., & Nidetzky, B. (1998). Structural and functional properties of a yeast xylitol dehydrogenase, a Zn2+-containing metalloenzyme similar to medium-chain sorbitol dehydrogenases. *Biochemical Journal, 336*(1), 91–99.
82. Ditzelmüller, G., Kubicek, C. P., Wöhrer, W., & Röhr, M. (1984). Xylitol dehydrogenase from *Pachysolen tannophilus. FEMS Microbiology Letters, 25*(2–3), 195–198.
83. Rizzi, M., Harwart, K., Erlemann, P., Bui-Thanh, N. A., & Dellweg, H. (1989). Purification and properties of the NAD+-xylitol-dehydrogenase from the yeast *Pichia stipitis. Journal of Fermentation and Bioengineering, 67*(1), 20–24.
84. Rizzi, M., Harwart, K., Bui-Thanh, N. A., & Dellweg, H. (1989). A kinetic study of the NAD+-xylitol-dehydrogenase from the yeast *Pichia stipitis. Journal of Fermentation and Bioengineering, 67*(1), 25–30.
85. Du Preez, J. C. (1994). Process parameters and environmental factors affecting D-xylose fermentation by yeasts. *Enzyme and Microbial Technology, 16*(11), 944–956.
86. Bruinenberg, P. M., Van Dijken, J. P., & Scheffers, W. A. (1983). A theoretical analysis of NADPH production and consumption in yeasts. *Microbiology, 129*(4), 953–964.
87. Nissen, T. L., Anderlund, M., Nielsen, J., Villadsen, J., & Kielland-Brandt, M. C. (2001). Expression of a cytoplasmic transhydrogenase in *Saccharomyces cerevisiae* results in formation of 2-oxoglutarate due to depletion of the NADPH pool. *Yeast, 18*(1), 19–32.
88. Anderlund, M., Rådström, P., & Hahn-Hägerdal, B. (2001). Expression of bifunctional enzymes with xylose reductase and xylitol dehydrogenase activity in *Saccharomyces cerevisiae* alters product formation during xylose fermentation. *Metabolic Engineering, 3*(3), 226–235.
89. Bicho, P. A., Runnals, P. L., Cunningham, J. D., & Lee, H. (1988). Induction of xylose reductase and xylitol dehydrogenase activities in *Pachysolen tannophilus* and Pichia stipitis on mixed sugars. *Applied and Environmental Microbiology, 54*(1), 50–54.
90. Saleh, A. A., Watanabe, S., Annaluru, N., Kodaki, T., & Makino, K. (2006, November). Construction of various mutants of xylose metabolizing enzymes for efficient conversion of biomass to ethanol. In *Nucleic acids symposium series* (Vol. 50, No. 1, pp. 279–280). Oxford University Press.
91. Kostrzynska, M., Sopher, C. R., & Lee, H. (1998). Mutational analysis of the role of the conserved lysine-270 in the *Pichia stipitis* xylose reductase. *FEMS Microbiology Letters, 159*(1), 107–112.

92. Wang, P. Y., & Schneider, H. (1980). Growth of yeasts on D-xylulose. *Canadian Journal of Microbiology, 26*(9), 1165–1168.
93. Yu, S., Jeppsson, H., & Hahn-Hägerdal, B. (1995). Xylulose fermentation by Saccharomyces cerevisiae and xylose-fermenting yeast strains. *Applied Microbiology and Biotechnology, 44*(3), 314–320.
94. Chang, S. F., & Ho, N. W. (1988). Cloning the yeast xylulokinase gene for the improvement of xylose fermentation. *Applied Biochemistry and Biotechnology, 17*(1), 313–318.
95. Ho, N. W., Chen, Z., & Brainard, A. P. (1998). Genetically engineered Saccharomyces yeast capable of effective cofermentation of glucose and xylose. *Applied and Environmental Microbiology, 64*(5), 1852–1859.
96. Ligthelm, M. E., Prior, B. A., du Preez, J. C., & Brandt, V. (1988). An investigation of D-{1-13C} xylose metabolism in *Pichia stipitis* under aerobic and anaerobic conditions. *Applied Microbiology and Biotechnology, 28*(3), 293–296.
97. Senac, T., & Hahn-Hägerdal, B. (1990). Intermediary metabolite concentrations in xylulose- and glucose-fermenting *Saccharomyces cerevisiae* cells. *Applied and Environmental Microbiology, 56*(1), 120–126.
98. Martín, C., & Jönsson, L. J. (2003). Comparison of the resistance of industrial and laboratory strains of Saccharomyces and Zygosaccharomyces to lignocellulose-derived fermentation inhibitors. *Enzyme and Microbial Technology, 32*(3-4), 386–395.
99. Brandberg, T., FranzéN, C. J., & Gustafsson, L. (2004). The fermentation performance of nine strains of *Saccharomyces cerevisiae* in batch and fed-batch cultures in dilute-acid wood hydrolysate. *Journal of Bioscience and Bioengineering, 98*(2), 122–125.
100. Zaldivar, J., Borges, A., Johansson, B., Smits, H., Villas-Bôas, S., Nielsen, J., & Olsson, L. (2002). Fermentation performance and intracellular metabolite patterns in laboratory and industrial xylose-fermenting *Saccharomyces cerevisiae. Applied Microbiology and Biotechnology, 59*(4), 436–442.
101. Sonderegger, M., Jeppsson, M., Larsson, C., Gorwa-Grauslund, M. F., Boles, E., Olsson, L., Spencer-Martins, I., Hahn-Hägerdal, B., & Sauer, U. (2004). Fermentation performance of engineered and evolved xylose-fermenting *Saccharomyces cerevisiae* strains. *Biotechnology and bioengineering, 87*(1), 90–98.
102. Olsson, L., Soerensen, H. R., Dam, B. P., Christensen, H., Krogh, K. M., & Meyer, A. S. (2006). Separate and simultaneous enzymatic hydrolysis and fermentation of wheat hemicellulose with recombinant xylose utilizing *Saccharomyces cerevisiae.* In *Twenty-Seventh Symposium on Biotechnology for Fuels and Chemicals* (pp. 117–129). Humana Press.
103. Matsushika, A., Inoue, H., Murakami, K., Takimura, O., & Sawayama, S. (2009). Bioethanol production performance of five recombinant strains of laboratory and industrial xylose-fermenting *Saccharomyces cerevisiae. Bioresource Technology, 100*(8), 2392–2398.
104. Garay-Arroyo, A., Covarrubias, A. A., Clark, I., Nino, I., Gosset, G., & Martinez, A. (2004). Response to different environmental stress conditions of industrial and laboratory *Saccharomyces cerevisiae* strains. *Applied Microbiology and Biotechnology, 63*(6), 734–741.
105. Nilsson, A., Gorwa-Grauslund, M. F., Hahn-Hägerdal, B., & Lidén, G. (2005). Cofactor dependence in furan reduction by *Saccharomyces cerevisiae* in fermentation of acid-hydrolyzed lignocellulose. *Applied and Environmental Microbiology, 71*(12), 7866–7871.
106. Eliasson, A., Boles, E., Johansson, B., Österberg, M., Thevelein, J. M., Spencer-Martins, I., Juhnke, H., & Hahn-Hägerdal, B. (2000). Xylulose fermentation by mutant and wild-type strains of Zygosaccharomyces and *Saccharomyces cerevisiae. Applied Microbiology and Biotechnology, 53*(4), 376–382.
107. Johansson, B., & Hahn-Hägerdal, B. (2002). The non-oxidative pentose phosphate pathway controls the fermentation rate of xylulose but not of xylose in *Saccharomyces cerevisiae* TMB3001. *FEMS Yeast Research, 2*(3), 277–282.
108. Matsushika, A., Inoue, H., Watanabe, S., Kodaki, T., Makino, K., & Sawayama, S. (2009). Efficient bioethanol production by a recombinant flocculent *Saccharomyces cerevisiae* strain with a genome-integrated NADP+-dependent xylitol dehydrogenase gene. *Applied and Environmental Microbiology, 75*(11), 3818–3822.

109. Sonderegger, M., & Sauer, U. (2003). Evolutionary engineering of *Saccharomyces cerevisiae* for anaerobic growth on xylose. *Applied and Environmental Microbiology, 69*(4), 1990–1998.
110. Jin, Y. S., Laplaza, J. M., & Jeffries, T. W. (2004). *Saccharomyces cerevisiae* engineered for xylose metabolism exhibits a respiratory response. *Applied and Environmental Microbiology, 70*(11), 6816–6825.
111. Attfield, P. V., & Bell, P. J. (2006). Use of population genetics to derive nonrecombinant *Saccharomyces cerevisiae* strains that grow using xylose as a sole carbon source. *FEMS Yeast Research, 6*(6), 862–868.
112. https://www.semanticscholar.org/paper/Development-on-ethanol-production-from-xylose-by-Yang-Lu/2f36879b17504762f0d5470ccb9cc6a90e5e8987/figure/1
113. https://www.semanticscholar.org/paper/Genetic-improvement-of-Saccharomyces-cerevisiae-for-Chu-Lee/4790ba9fb6d29584df72d891c460fa3bf9b3ae41/figure/0

Chapter 13
Prospects of Nanotechnology in Bioethanol Production

Abstract Converting lignocellulosic material to fermentable sugar is a complex process that is also the most expensive portion of the biofuel production process. Even though acid/enzyme catalysts are commonly employed for lignocellulosic material hydrolysis, enzyme immobilization is now being recognized as a viable option. Immobilization of various biocatalysts, such as cellulase, β-glucosidase, cellobiose, xylanase, laccase, and others, on support materials, such as nanomaterials, to generate nano biocatalysts improves catalytic efficiency and enzymatic stability. The use of nano biocatalysts boosts enzymatic stability and enzymatic hydrolysis efficiency. Various nanoparticles and nanomaterials have been discovered to have an impact on the synthesis of biofuels such as bioethanol. In this chapter, we look at some of the more attractive techniques and recent developments in the application of nanotechnology to the production of bioethanol.

Keywords Nanobiotechnology · Lignocellulose · Hydrolysis · Pretreatment · Bioethanol

13.1 Introduction

Several attempts have been made in recent decades to find and use a variety of materials for bioethanol production. Because substrate compatibility is one of the most important cost factors considered in large-scale ethanol production, it is preferable to use lignocellulosic biomass for both economic and environmental reasons. As a result, ethanol must be produced utilizing low-cost, carbohydrate-rich feedstocks [1]. Different techniques are used to make bioethanol from different bioresources (crops and lignocellulosic biomass) in different parts of the world, depending on where they can be found [2, 3]. Various physical, chemical, physicochemical, and biological approaches have been used for pretreatment. However, these approaches are costly and produce harmful inhibitors [4]. These concerns have been addressed by the development of nanotechnology, a new cost-effective, productive, and environmentally friendly process that offers a viable alternative to traditional pretreatment procedures. The nanotechnology method is widely used in a variety of industries, including biofuel production [5]. Nanotechnology has grown rapidly in recent

N. K. Aggarwal et al., *Bioethanol Production*, Green Chemistry and Sustainable Technology, https://doi.org/10.1007/978-3-031-05091-6_13

years around the world, bringing significant discoveries and benefits to an increasing variety of goods in fields as diverse as biotechnology, energy, environment, health, agriculture, and food [6]. In recent years, nanotechnology has grown to encompass nearly every field of science, with various applications in everyday life. In nanoscale material processes, nanomaterials are used [7].

13.2 Nanomaterials

Researchers are interested in using nanocatalysts in the production of bio-alcohols because of their reusability [8]. Nanomaterials are structures with a size varying from 1 to 100 nm (nm). They can be made in a variety of shapes and sizes [9]. Catalysts can be made from a variety of materials, including cobalt and nickel, as well as metal oxides [10, 11]. Magnetic nanoparticles (MNPs) have considerable promise for biofuel and bioenergy applications, such as the immobilization of enzymes like cellulases and hemicellulases on magnetic supports for the production of sugars and bioethanol from lignocellulosic biomass. It is possible to magnetically retrieve and recycle these immobilized enzymes for reuse [12]. Arc discharge, laser ablation, and chemical vapor deposition are all ways of producing carbon nanotubes (CNTs). These particles are made up of rolled-up graphite sheets with a nanometric diameter and great biocompatibility, making them a popular choice for enzyme immobilization [13]. Researchers have been very interested in CNTs conjugated with enzymes in recent years because of their potential biotechnological application in the areas of biosensor creation and biofuel generation [14]. In addition to magnetic ones, silica nanoparticles can also be employed to immobilize lignocellulolytic enzymes such as cellulase. In simultaneous saccharification and fermentation (SSF) processes, Lupoi and Smith demonstrated the catalytic activity of cellulase immobilized on silica-nanoparticles for increased production of bioethanol from lignocellulosic biomass [15]. It was discovered that silica-nanoparticles have a higher affinity for cellulase adsorption. When compared to unbound cellulase, immobilized cellulase produced 1.6 times the quantity of sugars from cellulose. Furthermore, SSF reactions involve the simultaneous hydrolysis of cellulose to glucose by cellulase as well as the simultaneous fermentation of glucose to ethanol by yeast. Furthermore, the yield of bioethanol was found to be enhanced by 2.1–2.3 times as a result of using the SSF technique.

13.3 Pretreatment

Pretreatment, which helps to alter the strong structure of lignocellulosic biomass and therefore enhances the accessibility of cellulases toward polysaccharides, allowing them to hydrolyze them into monosaccharides, is the most important stage in the synthesis of ethanol. It has also been suggested that pretreatment accounts for more than 40% of total operational costs. To fractionate, solubilize, hydrolyze, and separate cellulose, hemicellulose, and lignin components, a variety of pretreatment technologies are currently available. Pretreatment of lignocellulosic materials generally breaks down the complex structure, allowing fermentable sugars to be released by enzymatic saccharification [16]. The pretreatment of lignocellulose has traditionally included a wide range of physical, chemical, physicochemical, and biological techniques. Physical pretreatment reduces the particle size of biomass, which improves enzymatic accessibility [17]. Similarly, chemical pretreatment successfully changes the lignocellulosic crystalline structure to an amorphous form without requiring additional energy [18]. Nanomaterials of a smaller size can easily enter the cell walls of lignocellulosic materials and react with lignocellulosic components (cellulose and hemicellulose) to liberate monomeric and oligomeric sugars [19]. Pretreatment nanomaterials can improve the chemistry of organic matter at the molecular level. Some recent studies have shown that nanotechnology can provide simple, green, productive, and commercially sustainable lignocellulose pretreatment methods [7]. Pretreatment and enzyme hydrolysis stages account for a large portion of the cost of converting biomass to bioethanol/biochemicals. It is because enzymes are expensive, and if they can be recovered and repurposed with the help of nanoparticles, the bioprocess's economics will improve. Both of these stages use a nanotechnological technique to enhance cellulose conversion [20–22]. The fabrication and characterization of a sulfonated silica-magnetic-nanoparticle composite as a catalyst in the digestion of lignocellulose were reported by Lai et al. [23]. These nanocomposites are made up of sulfonated mesoporous silica as a capable hydrolysis catalyst that has been enhanced with magnetic iron oxide nanoparticles. According to the authors, this type of nanocatalyst could help with lignocellulose hydrolysis problems. Pena et al. [24] confirmed the catalytic efficacy of nanocatalysts for the processing of corn stover, specifically propyl-sulfonic-acid-functionalized-nanoparticles. To make this nanocatalyst, silica-coated-nanoparticles were functionalized in a solution of H_2O and C_2H_5OH containing 0.5% mercaptopropyltrimethoxysilane (MPTMS) at a neutral pH. The efficiency of these propyl-sulfonic-acid-functionalized-nanoparticles was evaluated at three distinct catalyst loads: 0.1, 0.2, and 0.3 g of catalyst per g of biomass, for one hour at three different temperatures: 160, 180, and 200 °C. The results showed that the catalyst load did not affect sugar yield at 160 °C, with an average sugar yield of approximately 59.0 percent produced from corn-stover hydrolysis at this temperature. Furthermore, samples incubated at 180 °C with a catalyst load of 0.2 g yielded a maximum sugar yield of 90%. At 200 °C, however, full-sugar hydrolysis was found. Wang et al. [25] described a rapid and efficient NSHA (Nanoscale shear hybrid alkaline) treatment strategy for biofuel production using corn stover as a lignocellulosic material. In this experiment, corn-stover cellulose and NaOH were mixed in a 1:1 ratio at room temperature. This method made it easier to remove a large amount of lignin and hemicellulose before disrupting cellulose in nanoform. Because of the NSHA pretreatment, the saccharification and total conversion rates of holocellulose were significantly enhanced.

13.4 Enzymatic Hydrolysis

Lignocellulosic hydrolysis, which turns biomass-derived cellulose and hemicellulose into simple sugars, is one of the most important steps in the process of making bioethanol. Enzymatic hydrolysis of lignocellulose is considered to be a more efficient approach than chemical hydrolysis. Enzymes such as cellulases, hemicellulases, and laccases are used to hydrolyze lignocellulosic biomass. The hydrolysis of cellulose necessitates the use of a cocktail of enzymes known as cellulases, which work together to convert cellulose into fermentable sugars. Enzymatic cellulose hydrolysis is a crucial step in biochemical conversion because it dictates the quality and quantity of the final product, which has an impact on downstream processing. Endo-β-1,4-glucanases, Exo-β-1,4-glucanases (cellobiohydrolase), and β-glucosidases are the three enzyme systems that makeup cellulase. Endocellulase enzymes sever β-1, 4 internal connections at random, resulting in free chain ends. Exocellulases also create disaccharides like cellobiose by acting on 2–4 units at the ends of exposed chains. Finally, cellobiose units are hydrolyzed by β-glucosidase, yielding fermentable sugars [26]. Several additional enzymes, like xylanase, β-xylosidase, glucomannase, and others, are also necessary for hemicellulose breakdown. This set of enzymes has the ability to hydrolyze both cellulose and hemicellulose components [27]. Even though enzyme-mediated lignocellulosic hydrolysis is an effective approach for biomass hydrolysis, the total process is costly due to the expensive enzyme production step. Compared to acid hydrolysis, enzymatic hydrolysis has a better degree of selectivity and does not produce inhibitor chemicals. Because enzyme production is so expensive, it accounts for the bulk of this rise. Biocatalysts for lignocellulosic hydrolysis have recently been developed using nanoparticles, nanocomposites, and nanomaterials. For enzyme hydrolysis, researchers have used nanomaterials to immobilize enzymes such as cellulase, hemicellulase, and ligninolytic. Nanobiocatalyst immobilized functionalized nanoparticles may be a game-changing alternative to established techniques since they may be retrieved and reused in hydrolytic reactions [28]. Many nanobiocatalysts have been used in the hydrolysis of biomass, and for enzyme immobilization, several nanomaterials could be used [5]. Several biocatalysts, including β-glucosidase, cellulase, laccase, cellobiose, and xylanase, are immobilized on nanomaterial supports to generate nanobiocatalysts, which improve the enzymes' catalytic potency and durability. Jordan et al. [29] successfully fixed the cellulase enzyme complex onto magnetic nanoparticles less than 1 μm in size using carbodiimide activation, which can then be used to treat cellulosic materials during ethanol production. Furthermore, the immobilization of enzymes on magnetic nanoparticles for the manufacture of magnetic nanobiocatalysts is preferred because these nanobiocatalysts may be easily recovered or reused in large-scale production parameters (Fig. 13.1).

Because nanomaterials have more surface area for enzyme binding, enzymes that are immobilized on them could function better, which could lead to more enzyme loading per unit mass of particles [30]. Efficient biofuel and fermentable sugar production could be achieved using nanoparticle-enhanced enzyme catalysis

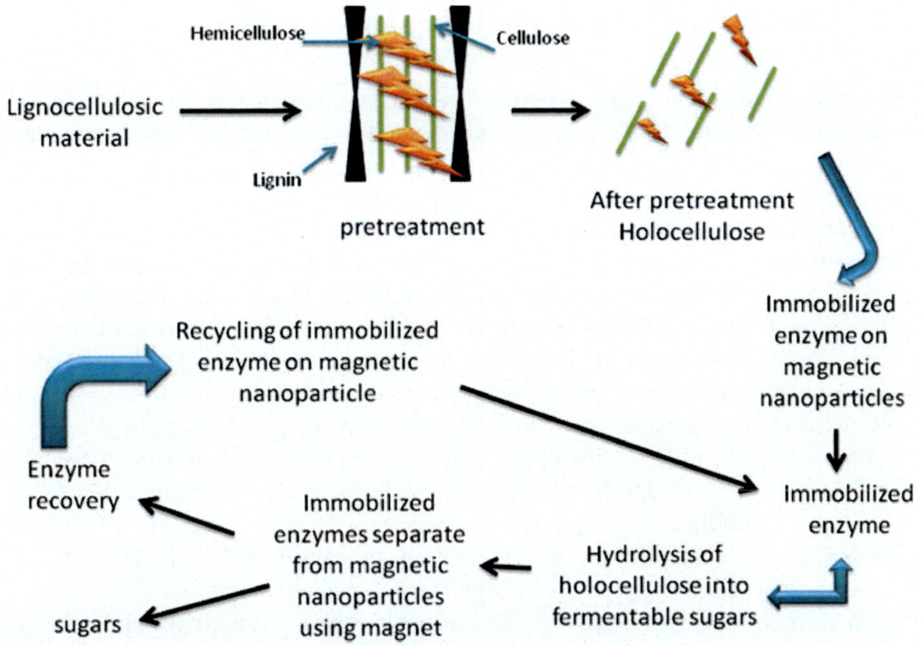

Fig. 13.1 Holocellulose hydrolysis using immobilized enzymes on magnetic nanoparticles and their recovery after hydrolysis into fermentable sugars from lignocellulosic materials

(i.e., by immobilizing holocellulose on magnetic nanoparticles), and the immobilized enzymes could be magnetically recovered and reused [31]. *Trichoderma reesei* cellulase immobilized on chitosan-associated magnetic nanoparticles was used to hydrolyze carboxymethylcellulose, and it retained over 80% hydrolytic activity even after 15 cycles [32] (Table 13.1).

Cellulases immobilized on ferrite NPs using the glutaraldehyde cross-linking agent hydrolyzed lignocellulosic biomass faster at 60 °C and remained active for the next three cycles [35]. A few investigations have shown that lignocellulosic biomass hydrolysis employing free enzymes converts cellulose to sugar with higher hydrolytic efficiency (78%) than cellulases immobilized on MNPs (72%) [38].

Table 13.1 Variety of nanomaterials that can be utilized to immobilize enzymes

Immobilized enzymes	Nanomaterials	References
Cellulase	Ag NPs and Au NPs	[33]
	MnO_2 NPs	[34]
	Zinc ferrite NPs	[35]
	Zn MNPs	[36]
	Fe_3O_4 MNPs	[29]
	Chitosan-coated MNPs	[32]
Xylanase	Mg NPs	[37]

13.5 Bioethanol Production Using Various Nanomaterials

It is becoming increasingly popular to use nanoparticles in the production of bioethanol because of their high interfacial surface area, which enhances biochemical processes' kinetics. In specific quantities, supplemental nanoparticles can boost fermentative ethanol production and change the microorganism's metabolic route to promote ethanol production [8, 39]. According to El-Kemary et al. [40], the hydrophobic component of $C_6H_{12}O_6$ was adsorbed onto the surface layers of nanoparticles by Vander Waals interactions, while the hydrophilic component of the OH is oriented toward the aqueous phase. As this study found, the reaction between the nanoparticles and the sugar made it easier for *S. cerevisiae* BY4743 to take in glucose, which led to more bioethanol production. Fe_3O_4 nanoparticles significantly increased ethanol production, with a highest bioethanol yield of 0.26 g/g, hexose sugar consumption of 99.95%, ethanol productivity of 0.22 g/L/h, and fermentation efficiency of 51% at 0.01% wt [41]. For *Sesbania aculeate* cellulose hydrolysis, cellulase immobilized on magnetic nanoparticles was employed to yield 5.31 g/L of ethanol, and the nanobiocatalysts were reused several times [42] (Table 13.2).

According to Ivanova et al. [46], alginate and magnetic nanoparticles (entrapped yeast cells) covalently bonded to chitosan-magnetite microparticles and cellulose-coated MNPs were shown to increase bioethanol fermentation. With yeast cells encased in a matrix of alginate/magnetic nanoparticles and immobilized on magnetite-containing chitosan, the ethanol output was reported to be about 91% of the theoretical yield. At 4 °C in saline, the self-life of these MNPs was reported to be more than one month. For bioethanol production, cellulases produced by Aspergillus fumigatus were immobilized on manganese oxide nanoparticles, resulting in binding efficiency of 75%. These immobilized cellulases were reused for five cycles and demonstrated enhanced stability over a wide range of temperatures and pH. Finally, employing free and immobilized cellulase, bioethanol production from sugarcane leaves yielded 18 and 22 g/L of bioethanol, respectively [34].

Table 13.2 Nanomaterials for bioethanol productions

S. No.	Nanomaterials	Ethanol yield	References
1	ZnO NPs	0.0193 g/g	[43]
2	Fe_3O_4 NPs	0.26 g/g	[41]
3	ZnO NPs	0.0359 g/g	[44]
4	Methyl-functionalized silica NPs	0.306 ± 0.0028 g/L	[45]
5	Entrapped yeast cells MNPs	264.0 g/L	[46]

13.6 Summary

A Nanobiotechnology technique could be used as a possible pretreatment for lignocellulosic material conversion. The hydrolysis of lignocellulosic materials is an important stage in the biofuel generation process. Acid hydrolysis processes result in the production of harmful inhibitors, which necessitates time-consuming downstream processing. The hydrolysis process strategy is exceedingly selective and environmentally safe. However, the process is expensive due to high-cost enzymes. The stability and proficiency of enzymes are increased by immobilizing them on various nanoparticles. Furthermore, immobilizing enzymes on magnetic nanoparticles produces nanobiocatalysts with strong potential for cellulose hydrolysis, which is well-known. The use of nanomaterials such as nanocomposites, MNPs, nanocatalysts, and nanotubes, as well as processes such as the NSHA Technique, has helped to improve the economic viability of the production process. Nanomaterials have the potential to play a key role in accomplishing the goal of commercial-scale bioethanol production at a low cost, and their use is expected to grow as the field of fermentation-based bioethanol production progresses. Nanotechnology has improved the longevity of the bioethanol manufacturing process by allowing many of its nanoparticles to be reused. The application of such nanotechnological developments in the future may open up new paths for the long-term bioethanol production.

References

1. Talasila, U., & Vechalapu, R. R. (2015). Optimization of medium constituents for the production of bioethanol from cashew apple juice using Doehlert experimental design. *International Journal of Fruit Science, 15*(2), 161–172.
2. Shanavas, S., Padmaja, G., Moorthy, S. N., Sajeev, M. S., & Sheriff, J. T. (2011). Process optimization for bioethanol production from cassava starch using novel eco-friendly enzymes. *Biomass and Bioenergy, 35*(2), 901–909.
3. Puri, M., Abraham, R. E., & Barrow, C. J. (2012). Biofuel production: Prospects, challenges and feedstock in Australia. *Renewable and Sustainable Energy Reviews, 16*(8), 6022–6031.
4. Kumar, A. K., & Sharma, S. (2017). Recent updates on different methods of pretreatment of lignocellulosic feedstocks: A review. *Bioresources and bioprocessing, 4*(1), 1–19.
5. Rai, M., Ingle, A. P., Pandit, R., Paralikar, P., Biswas, J. K., & da Silva, S. S. (2019). Emerging role of nanobiocatalysts in hydrolysis of lignocellulosic biomass leading to sustainable bioethanol production. *Catalysis Reviews, 61*(1), 1–26.
6. Pérez-López, B., & Merkoçi, A. (2012). Carbon nanotubes and graphene in analytical sciences. *Microchimica Acta, 179*(1), 1–16.
7. Rai, M., dos Santos, J. C., Soler, M. F., Marcelino, P. R. F., Brumano, L. P., Ingle, A. P., Gaikwad, S., Gade, A., & da Silva, S. S. (2016). Strategic role of nanotechnology for production of bioethanol and biodiesel. *Nanotechnology Reviews, 5*(2), 231–250.
8. Kim, Y. K., & Lee, H. (2016). Use of magnetic nanoparticles to enhance bioethanol production in syngas fermentation. *Bioresource Technology, 204*, 139–144.
9. Tiwari, J. N., Tiwari, R. N., & Kim, K. S. (2012). Zero-dimensional, one-dimensional, two-dimensional and three-dimensional nanostructured materials for advanced electrochemical energy devices. *Progress in Materials Science, 57*(4), 724–803.

10. Kodama, R. H. (1999). Magnetic nanoparticles. *Journal of Magnetism and Magnetic Materials, 200*(1–3), 359–372.
11. Laurent, S., Forge, D., Port, M., Roch, A., Robic, C., Vander Elst, L., & Muller, R. N. (2008). Magnetic iron oxide nanoparticles: Synthesis, stabilization, vectorization, physicochemical characterizations, and biological applications. *Chemical Reviews, 108*(6), 2064–2110.
12. Alftrén, J., & Hobley, T. J. (2013). Covalent immobilization of β-glucosidase on magnetic particles for lignocellulose hydrolysis. *Applied Biochemistry and Biotechnology, 169*(7), 2076–2087.
13. Feng, W., & Ji, P. (2011). Enzymes immobilized on carbon nanotubes. *Biotechnology Advances, 29*(6), 889–895.
14. Shi, Q., Yang, D., Su, Y., Li, J., Jiang, Z., Jiang, Y., & Yuan, W. (2007). Covalent functionalization of multi-walled carbon nanotubes by lipase. *Journal of Nanoparticle Research, 9*(6), 1205–1210.
15. Lupoi, J. S., & Smith, E. A. (2011). Evaluation of nanoparticle-immobilized cellulase for improved ethanol yield in simultaneous saccharification and fermentation reactions. *Biotechnology and Bioengineering, 108*(12), 2835–2843.
16. Takano, M., & Hoshino, K. (2018). Bioethanol production from rice straw by simultaneous saccharification and fermentation with statistical optimized cellulase cocktail and fermenting fungus. *Bioresources and Bioprocessing, 5*(1), 1–12.
17. Madadi, M., Tu, Y., & Abbas, A. (2017). Pretreatment of lignocelollusic biomass based on improving enzymatic hydrolysis. *International Journal of Applied Sciences and Biotechnology, 5*(1), 1–11.
18. Chen, W. H., Chen, Y. C., & Lin, J. G. (2014). Study of chemical pretreatment and enzymatic saccharification for producing fermentable sugars from rice straw. *Bioprocess and Biosystems Engineering, 37*(7), 1337–1344.
19. Rai, M., Ingle, A. P., Gaikwad, S., Dussán, K. J., & Silva, S. S. D. (2017). Role of nanoparticles in enzymatic hydrolysis of lignocellulose in ethanol. In *Nanotechnology for bioenergy and biofuel production* (pp. 153–171). Springer, Cham.
20. Verma, M. L., & Barrow, C. J. (2015). Recent advances in feedstocks and enzyme-immobilised technology for effective transesterification of lipids into biodiesel. *Microbial Factories*, 87–103.
21. Kumar, A., Sharma, S., Pandey, L. M., & Chandra, P. (2018). Nanoengineered material based biosensing electrodes for enzymatic biofuel cells applications. *Materials Science for Energy Technologies, 1*(1), 38–48.
22. Kumar, A., Purohit, B., Maurya, P. K., Pandey, L. M., & Chandra, P. (2019). Engineered nanomaterial assisted signal-amplification strategies for enhancing analytical performance of electrochemical biosensors. *Electroanalysis, 31*(9), 1615–1629.
23. Lai, D. M., Deng, L., Guo, Q. X., & Fu, Y. (2011). Hydrolysis of biomass by magnetic solid acid. *Energy & Environmental Science, 4*(9), 3552–3557.
24. Pe, L., Xu, F., Hohn, K. L., Li, J., & Wang, D. (2014). Propyl-sulfonic acid functionalized nanoparticles as catalyst for pretreatment of corn stover. *Journal of Biomaterials and Nanobiotechnology, 2014.*
25. Wang, W., Ji, S., & Lee, I. (2013). Fast and efficient nanoshear hybrid alkaline pretreatment of corn stover for biofuel and materials production. *Biomass and Bioenergy, 51*, 35–42.
26. Chaturvedi, V., & Verma, P. (2013). An overview of key pretreatment processes employed for bioconversion of lignocellulosic biomass into biofuels and value added products. *3 Biotech, 3*(5), 415–431.
27. Verardi, A., De Bari, I., Ricca, E., Calabrò, V. (2012). Hydrolysis of lignocellulosic biomass: current status of processes and technologies and future perspectives. In *Bioethanol* (Vol. 2012, pp. 95–122). Rijeka: InTech.
28. Zhang, W., Qiu, J., Feng, H., Zang, L., & Sakai, E. (2015). Increase in stability of cellulase immobilized on functionalized magnetic nanospheres. *Journal of Magnetism and Magnetic Materials, 375*, 117–123.
29. Jordan, J., Kumar, C. S., & Theegala, C. (2011). Preparation and characterization of cellulase-bound magnetite nanoparticles. *Journal of Molecular Catalysis B: Enzymatic, 68*(2), 139–146.

30. Grewal, J., Ahmad, R., & Khare, S. K. (2017). Development of cellulase-nanoconjugates with enhanced ionic liquid and thermal stability for in situ lignocellulose saccharification. *Bioresource Technology, 242*, 236–243.
31. Kim, K. H., Lee, O. K., & Lee, E. Y. (2018). Nano-immobilized biocatalysts for biodiesel production from renewable and sustainable resources. *Catalysts, 8*(2), 68.
32. Sanchez-Ramirez, J., Martinez-Hernandez, J. L., Segura-Ceniceros, P., Lopez, G., Saade, H., Medina-Morales, M. A., Ramos-González, R., Aguilar, C. N., & Ilyina, A. (2017). Cellulases immobilization on chitosan-coated magnetic nanoparticles: Application for Agave Atrovirens lignocellulosic biomass hydrolysis. *Bioprocess and Biosystems Engineering, 40*(1), 9–22.
33. Mishra, A., & Sardar, M. (2015). Cellulase assisted synthesis of nano-silver and gold: Application as immobilization matrix for biocatalysis. *International Journal of Biological Macromolecules, 77*, 105–113.
34. Cherian, E., Dharmendirakumar, M., & Baskar, G. (2015). Immobilization of cellulase onto MnO_2 nanoparticles for bioethanol production by enhanced hydrolysis of agricultural waste. *Chinese Journal of Catalysis, 36*(8), 1223–1229.
35. Manasa, P., Saroj, P., & Korrapati, N. (2017). Immobilization of cellulase enzyme on zinc ferrite nanoparticles in increasing enzymatic hydrolysis on ultrasound-assisted alkaline pretreated Crotalaria juncea biomass. *Indian Journal of Science and Technology, 10*(24), 1–7.
36. Abraham, R. E., Verma, M. L., Barrow, C. J., & Puri, M. (2014). Suitability of magnetic nanoparticle immobilised cellulases in enhancing enzymatic saccharification of pretreated hemp biomass. *Biotechnology for Biofuels, 7*(1), 1–12.
37. Dutta, N., & Saha, M. K. (2019). Nanoparticle-induced enzyme pretreatment method for increased glucose production from lignocellulosic biomass under cold conditions. *Journal of the Science of Food and Agriculture, 99*(2), 767–780.
38. Yang, B., & Wyman, C. E. (2004). Effect of xylan and lignin removal by batch and flowthrough pretreatment on the enzymatic digestibility of corn stover cellulose. *Biotechnology and Bioengineering, 86*(1), 88–98.
39. Han, H., Cui, M., Wei, L., Yang, H., & Shen, J. (2011). Enhancement effect of hematite nanoparticles on fermentative hydrogen production. *Bioresource Technology, 102*(17), 7903–7909.
40. El-Kemary, M., Nagy, N., & El-Mehasseb, I. (2013). Nickel oxide nanoparticles: Synthesis and spectral studies of interactions with glucose. *Materials Science in Semiconductor Processing, 16*(6), 1747–1752.
41. Sanusi, I. A., Faloye, F. D., & Gueguim Kana, E. B. (2019). Impact of various metallic oxide nanoparticles on ethanol production by Saccharomyces cerevisiae BY4743: Screening, kinetic study and validation on potato waste. *Catalysis Letters, 149*(7), 2015–2031.
42. Baskar, G., Kumar, R. N., Melvin, X. H., Aiswarya, R., & Soumya, S. (2016). Sesbania aculeate biomass hydrolysis using magnetic nanobiocomposite of cellulase for bioethanol production. *Renewable Energy, 98*, 23–28.
43. Zada, B., Mahmood, T., Malik, S. A., & Uddin, Z. (2014). Effect of zinc oxide nanoparticles on hyacinth's fermentation. *International Journal for Enhancements Research Science Technology Engineering, 3*, 78–92.
44. Gupta, K., & Chundawat, T. S. (2020). Zinc oxide nanoparticles synthesized using Fusarium oxysporum to enhance bioethanol production from rice-straw. *Biomass and Bioenergy, 143*, 105840.
45. Kim, Y. K., Park, S. E., Lee, H., & Yun, J. Y. (2014). Enhancement of bioethanol production in syngas fermentation with Clostridium ljungdahlii using nanoparticles. *Bioresource Technology, 159*, 446–450.
46. Ivanova, V., Petrova, P., & Hristov, J. (2011). Application in the ethanol fermentation of immobilized yeast cells in matrix of alginate/magnetic nanoparticles, on chitosan-magnetite microparticles and cellulose-coated magnetic nanoparticles. arXiv preprint arXiv:1105.0619.

Chapter 14
Life Cycle Analysis (LCA) in GHG Emission and Techno-economic Analysis (TEA) of Bioethanol Production

Abstract The exhaustion of fossil resources and poor conditions of the environment are two of the world's primary worries these days. Fossil fuels, like coal, oil, and natural gas, are being exploited at a rising rate in both developed and developing countries, resulting in their overall depletion. There is an urgent need to produce sustainable and green technologies. Bioethanol, liquid renewable biofuel is an ideal option to opt. Production at economical price keeping in view all the factors is very important. To assist the commercialization of enhanced ethanol production, each case should undergo a techno-economic analysis (TEA). Life cycle assessment is a strong technique for identifying and assessing environmental consequences, energy usage, and financial feasibility of alternative lignocellulosic ethanol pathways, as well as identifying sites for future revolution. Techno-economic analysis and Life cycle assessment provide a thorough foundation for decision-making. This chapter aims at concept of TEA and LCA of bioethanol production from different feedstocks.

Keywords Life cycle assessment · Techno-economic analysis · Bioethanol · Cost · Bioenergy · Environment

14.1 Introduction

Rising concern about the impact of climate change has prompted researchers to look into cleaner energy sources. Governments around the world, including the United Kingdom, are committed to achieving the Paris Agreement's climate change goals [1]. To achieve these objectives, infrastructure must be built on a foundation of sustainable phenomenon, not just to cut gaseous emissions but also to ameliorate the effects of existing environmental alterations, like long-lived carbon dioxide (CO_2) emissions. Transportation is a sector that is a key contributor to changes in climatic conditions, accounting for around 24% of worldwide carbon dioxide emissions and 21% of total GHG emissions; road vehicles account for three-quarters of these emissions [2, 3]. The burning of fossil fuels, primarily gasoline, coal, and diesel, emits these pollutants, as well as other greenhouse gases (GHGs). Furthermore, fossil fuels are a finite resource that is rapidly depleting. Green energy sources are being considered as a viable alternative source of energy for powering automobiles as part

N. K. Aggarwal et al., *Bioethanol Production*, Green Chemistry and Sustainable Technology, https://doi.org/10.1007/978-3-031-05091-6_14

of efforts to minimize GHG emissions. Hydrogen and renewable power are strong contenders, as are biofuels like bioethanol, biodiesel, and biogas [4]. Many vehicles, including aircraft, long-distance commuter vehicles, watercraft, and heavy-duty vehicles, are not yet capable of running on hydrogen due to logistical and cost issues [5, 6], while electric power is not yet viable in many vehicles, like in watercraft, aircraft, long-distance commuter vehicles, and heavy-duty vehicles. This is mostly owing to practical limitations in the spectrum of battery energy density [7], payload, and total cost. Biofuels, on the other hand, are both promising and readily usable alternative fuels. Biofuels have the potential to offset the carbon emitted during their manufacture.

Bioethanol is that a liquid renewable biofuel that is commonly blended with gasoline to boost the vehicle's octane rating and lower its carbon footprint. It is derived from sugars extracted from biological materials like sweet and starchy crops (first-generation bioethanol), lignocellulosic feedstocks (second-generation bioethanol), and seaweed (third-generation bioethanol). These biological materials collect CO_2 from the atmosphere through photosynthetic growth during culture. Fermentation of the resulting sugars also produces a very pure CO_2 stream, which can be used in carbon capture and storage (CCS) technology [8].

14.2 Life Cycle Analysis for Bioethanol Production

The Life Cycle Analysis is the most frequent tool for evaluating a process or product's environmental performance (LCA). The LCA research includes: to study the characteristics and description of product, analyzing the functional unit, system boundary, different categories which have impact on the product, and several assumptions or limitations [9]. Different studies are reported based on diverse range of substrates for production of bioethanol. LCA from different feedstocks is studied thoroughly. Some of the examples are given below.

Traditional bioethanol production is focused on inland sites and relies on edible crops and freshwater, resulting in food versus fuel controversy. So, this study established a solution by building a coastal marine biorefinery (CMB) system for production of bioethanol. The main goal of this study was to assess the environmental impact of using saline water at coastal sites for bioethanol production. As a result, an LCA was carried out using new feedstocks, rather than conventional ones. For comparison, the influence of each situation on climate change, water reduction, land usage, and fuel exhaustion was evaluated. In all of the identified impact categories, the Coastal Seawater scenario outperformed the conventional scenario. The usage of seawater in the procedure, in particular, had a considerable influence on H_2O depletion, with a decrease of 31.2 percent. Additionally, there were reductions of 5.5%, 3.5%, and 4.2% in natural land transformation, climatic alteration, and fossil reduction, respectively. This demonstrates the benefits of exploiting seawater and coastal regions for bioethanol production and stimulates further research into the CMB system [10].

Potential of shrub willow crop in northeastern region of US was carried out for LCA of bioethanol production [11]. Ethanol was generated from the fermentation of sugars from hot H_2O extract of willow planted on cropland liberated 0.0120.003 kg carbon dioxide eq MJ1. Despite the lower organic carbon levels in the soil, when willow is cultivated on grassland rather than in a field, the resulting fuel can still give significant climate advantages when compared to gasoline. Hydrolysis of sugars into bioethanol by using hot water extraction (HWE) is a promising method for lignocellulosic ethanol production. After different separation and purification steps, the HWE method can provide different chemicals like formic acid, acetic acid, furfural, lignin, and different fermentable sugars from woody biomass [12–15]. Hot water extraction is typically conducted without the utilization of any harsh chemicals at higher temperatures in the range of 140–180 °C for one to two h, which leads to hydrolysis of hemicellulose into H_2O [16]. The auto- hydrolysis of polymeric substances into monomeric sugar units. Extracted sugars derived from the extract can be transformed into different compounds like ethanol, butanol, or other value-added chemicals through fermentation. The residual HWE biomass has desirable characteristics like low ash content and high energy content, making it a useful feedstock for power and heat cogeneration [17]. In the northeast United States, the conversion of lignocellulosic biomass to bioethanol, high-yielding willow producers, and commercial level production presents new prospects for an emerging and thriving bioeconomy while offering many ecosystem benefits [18, 19]. There are few reports available on the LCA and greenhouse gas emissions of willow cultivation and conversion of ethanol [20, 21]. A study conducted a life cycle analysis (LCA) for ethanol synthesis by bioconversion of willow biomass, using steam explosion as a pretreatment method and enzymatic hydrolysis to convert cellulosic and hemicellulosic polymers into monomeric units prior to fermentation into final product [22]. Without considering the effects of land-use change, they observed harmful gaseous emission falls up to 120%, comparing ethanol from willow to gasoline. Researchers have studied the energy or environmental consequences of bioethanol production by biochemical method of conversion of willow in Sweden [20]. Simultaneous saccharification and fermentation in a willow biomass processing unit capable of processing 3000 Mg per day [23]. In comparison to gasoline, Stephenson and González-Garca discovered greenhouse gas savings of 70–90%. GHG reductions of 80% for concentrated hemicellulose, 68% for 5C sugars that strive with sugar derived from sugar cane, and greenhouse gas emission increase of 10% for concentrated hemicellulose (70% dry solid for animal feed) to replace molasses (72% dry solid) from sugar beet were found in an analysis of a HWE-based biorefinery combined with paper mills [24].

Residual waste of food has the potential to be used as bioethanol feedstocks, making them a feasible option for lowering greenhouse gas (GHG) emissions. Additionally, bioethanol synthesis from these wastes has the potential to reduce both waste disposal and cost of ethanol production [25] (Table 14.1).

The content of waste has a significant impact on the synthesis of bioethanol from it [34, 35]. The content of waste influences not only the end ethanol synthesis but also the particular procedures that contribute to the overall environmental impact. Pretreatment is the initial phase in the process, and it seeks to minimize particle

Table 14.1 LCA of different feedstocks for bioethanol production

Feedstock used	Output	Process	Key impacts	References
Organic waste	One ton of municipal wet biowaste, 1 MJ ethanol	Pretreatment, enzymatic hydrolysis, and fermentation	−15 kg carbon dioxide eq/ton biowaste as compared to available waste processing. Enhanced ethanol production was observed because of increase in enzyme loading	[26]
Municipal solid waste	1 MJ of ethanol one ton and municipal solid waste treated	Steam pretreatment, enzymatic saccharification, fermentation, and recovery of product	Greenhouse gas emissions lead – 566 g CO_2 eq/MJ bioethanol. Substantial contribution of downstream wastewater processing to greenhouse gas emissions. Due to excessive chemical usage during pretreatment, increased acidification was observed	[27]
Brewing waste	74.22 tons stream which is lignocellulosic in nature	Autohydrolysis pretreatment, purification, and fermentation	Steam production required higher temperature for autohydrolysis results in global warming. Production of enzyme was required in simultaneous saccharification and fermentation	[28]
LCB (derived from banana packaging setup)	One MJ of energy was liberated during combustion of ethanol	Simultaneous saccharification fermentation along with pretreatment through steam explosion	Increased acidification because of chemicals used in the process	[29]

(continued)

Table 14.1 (continued)

Feedstock used	Output	Process	Key impacts	References
Sugarcane	One ton of sugarcane	Juice extraction, hydrolysis, fermentation, distillation	All estimated cases showed positive values for climate change and fossil fuel reduction. It also reveals less efficiency in man toxicity	[30]
Municipal solid waste	One ton of waste is required	Selective degradation of cellulosic portion, and fermentation	Production of ethanol evidences to be the best alternative to avoid anthropological health and ecosystems impacts	[31]
Food processing waste	1-ton waste	Simultaneous saccharification and fermentation	Negative greenhouse gas emission and almost 500 percent enhancement as compared production from corn	[32]
Banana pulp, flower stalk, peel, and fruit	4000 kg of banana fruit per day	Diluted acid and enzymatic saccharification, and distillation	1.9 energy ratio for fruit and flesh assessed, which is comparatively higher than corn-based ethanol. Low energy is required when fruit was co-fermented with cellulose	[33]

size and enhance surface area. Domestic kitchen waste, for example, may include readily accessible carbohydrates for ethanol synthesis than waste obtained from industries. As a result, the pretreatment stage does not necessitate a lot of energy or additional ingredients like alkali, acid and organic solvents, etc. This is true for several high-sugar industrial residual food wastes also, such as palm oil fronds [36], banana waste [29], brewing waste [28], and fruit syrups [32], that can be used to make ethanol without requiring high energy pretreatment. The hydrolysis stage begins when the pretreatment step is completed. Enzymes, for example, amylases, cellulases, glycoamylases, hemicellulases) hydrolyze carbohydrates such as starch, cellulose, and hemicellulose into fermentable sugars at this stage. Enzyme synthesis consumes a lot of energy and steam [37, 38]. Production of glucoamylase has been proven to produce maximum harmful gaseous emissions when it comes to enzyme production [39, 40]. The power or energy requirements are responsible for the GHG emissions in glucoamylase manufacturing. Cellulase's effects, on the other hand, are linked to the chemicals and nutrients utilized during synthesis. Given the enzyme dosages utilized, the impact on the environment of cellulase in the generation of lignocellulosic ethanol is comparatively higher. Cellulase requires 25–250 times more dosages than amylase and glucoamylase [39]. According to González-Garca et al., enzymes and chemicals are responsible for 20% of GHG emissions produced during the LC of ethanol synthesized in a brewery waste system [28]. Due to the wide range of reported global warming impacts of enzymes, varied enzyme dosages, and production of ethanol, Literature estimated that enzyme role within GWI of biorefineries generating ethanol ranges from 11 to 62% [26]. Producing the requisite enzymes rather than using commercially available ones could be a way to cut costs and make lignocellulosic ethanol more affordable [41–43].

Separate hydrolysis and fermentation (SHF) and simultaneous saccharification and fermentation (SSF) are two alternative process setups for fermentative conversion to ethanol (SSF). SHF is a classic approach in which substrate is first hydrolyzed, then the monosaccharides are fermented into bioethanol in separate bioreactor. Hydrolysis and fermentation processes, on the other hand, can be joint in a procedure known as SSF. SHF has the benefit of allowing enzymes and microorganisms to work at their ideal levels, such as temperature. The growth of hydrolysis products, which limit action of enzymes and result in slower response rates, is the fundamental downside of SHF. Because the temperature in SSF is not suitable for activation of enzymes, the degree of hydrolysis is sluggish, however, SSF has several advantages over SHF: Lower production costs, less time consuming, small fermentor volumes, high yield of ethanol, reduced enzyme loads, and restricted enzymatic inhibition because of simultaneous removal of end product [44]. The environment hotspot in the profile of bioethanol generation by brewing waste, according to González-Garca et al., is connected with the SSF stage, owing to the usage of enzymes, which play a crucial role in all of the categories assessed [28]. To reduce the energy requirement during distillation, the ethanol content in the fermentation broth must be greater than 4% (w/w) [45, 46]. The quantity of sugars liberated in the digestion stage must be at least 80 g/L to meet this ethanol criterion. As a result, during enzymatic hydrolysis, solid loadings of more than 15% are required [47]. Due to mixing challenges, heat, and mass transport constraints, high solid loading causes increased viscosity. In high solids SSF, a fed-batch method could potentially mitigate these issues and lead to higher ethanol yields. The prehydrolysis and Simultaneous Saccharification and Fermentation (PSSF) process, which is a type of variation of the SSF method, in which the feedstock passes over a brief prehydrolysis step at optimum temp for enzymes followed by fermentation [48], is another way to overcome the high solids problem.

14.3 Techno-economic Analysis of Bioethanol Production

Both technical and economic factors determine the cost of ethanol production, including but not only restricted to bioconversion method and efficiency, waste treatment cost, logistic cost, feedstock type and price, government subsidy programs, and revenue from co-product sales [49, 50]. There are some vital factors that require to be considered on to lower down the production cost of ethanol—high conc. of ethanol at the time of fermentation, reduced cost of feedstock, high capacity of plant, maximum co-products liberation, lignin-based chemicals, etc. [51]. Advanced methods of pretreatment, enhanced enzymatic hydrolysis, usage of cost-effective and efficient enzymes, as well as improved fermentation process represent the dominant area or challenges in the field of bioethanol production. There are many methods employed for pretreatment, each of has its own advantage or disadvantage. This results in varied diversity in pretreatment cost as well as production of ethanol. Tables 14.2 and 14.3 describe about the direct capital of some pretreatment methods and operating cost of different pretreatment processes.

South African researchers [54] conducted an economic assessment for ethanol and electricity generation using combined techniques of first- and second-generation processes with sugarcane bagasse as substrates. He discovered that integrating cellulosic ethanol into present facilities like starch mills or autonomous distilleries (i.e., 1st generation ethanol plant) and converting on-site leftovers to biofuels, electricity, and heat might be cost-effective. Solitary 2nd generation plan, on the other hand, necessitates centralized facilities where LCB must be imported in order to generate biofuels and electrical energy. It has been proven that integrating these technologies

Table 14.2 Direct capital of some of the pretreatment methods [52]

Method	Corn stover $/(m^3.day)	Switchgrass $/(m^3.day)
Dilute acid	12,500	22,500
Ammonia fiber explosion	12,850	15,500
Liquid hot water	2,250	10,000
Lime	11,150	28,500
Steam explosion with SO_2	–	17,500

Table 14.3 Principal processing capital of different pretreatment processes [53]

Pretreatment	Principal processing cost
Liquid hot water	Energy consumption
Alkali extraction	Energy consumption, recovery of lime, and cost of lime
Steam explosion	Energy consumption
Direct sulfuric acid treatment	Energy consumption and cost of acid
Ammonia fiber explosion	Energy and cost of ammonia
Organosolvent method	Energy and solvent cost

reduces the MESP of 2nd generation ethanol. Second-generation ethanol production incorporated into an improved sugar mill had the lowest MESP ($0.38/L), followed by consolidated 1st—or 2nd generation ethanol production ($0.63/L) and second-generation freestanding ethanol synthesis ($0.66/L).

A study has investigated the commercial level synthesis of wheat straw-based ethanol on the Canadian Prairies in a study published in Canada. Yorkton was studied to be the most appropriate location in terms of production, out of 11 locations tested in Saskatchewan. The optimal plant size, based on economies of scale, was estimated to be around 250 million L annum-1. Feedstock cost was found to have a greater negative impact on smaller plants than on bigger plants. Within a substrate cost range of 35–70 $/t and irrespective of total capital venture, all plant capacities earned positive NPV at higher ethanol cost of 1.3–1.7 $/L. The high MESP in this study compared to gasoline prices recommended that government subsidies and R&D, like consolidated bioprocessing (CBP) and the supply chain of wheat straw, must be increased to make bioethanol more feasible in Canada [55].

Researchers explored ethanol synthesis from empty fruit bunches as a substrate for the palm oil extraction procedure in Brazil. The EFB 'feed-fuel' (EFB FF) case, which synthesized ethanol and syrup from pentose sugars as a cattle nutrient source, and the EFB 'only fuel' (EFB OF) scenario, which only synthesized ethanol as the principal product, were both investigated. Diluted acid and liquid hot water were used as pretreatment procedures for the EFB OF and EFB FF situations, respectively. The treated stream was then processed using the SHF method after pretreatment. EFB OF and EFB FF scenarios have ethanol production costs of $1038/t ($0.82/L) and $849/t ($0.67/L), respectively, compared to a market value of $760/t ($0.60/L). TCIs, VCs, and FCs for EFB OF and EFB FF cases were found to be roughly $226.6 and $204.1 million dollars, $8.3 million dollars per year and $7.9 million dollars per year, $8.9 million dollars per year, and $3.9 million dollars per year, respectively [56].

When compared to traditional ethanol synthesis from LCB in separate procedure, CBP (consolidated bioprocessing) based on a consortium of microorganisms operating at full-scale (2000 t/d) makes up to 27.5% of net ethanol production expenses. The price savings are primarily obtained through lower CAPEX because of the integrated process's reduced apparatus requirements, as well as reduced OPEX because glucose is not required for production of enzymes. When compared to literature estimates of CBP cost reductions based on genetically engineered microbes, the outputs are in the same vicinity. Yield and scale were recognized as the key cost-pushers from a procedure standpoint as a result of a rigorous sensitivity study, whereas the price level of the plant location has the greatest impact on the investment conditions. CBP provides enough edge for gainful production and the ability to decentralize biomass valorization in the EU, but in the world's prime bioethanol market, the United States, profitable production of lignocellulosic alcohol can only be obtained by combining CBP with other cost-cutting techniques, such as the use of cost-free waste feedstocks, because ethanol has experienced a significant price collapse [57].

In another report, bioethanol was generated from microwave-pretreated domestic kitchen waste. Pretreatment lasted 30 min at a continuous output of 90 watts. Amylase from Bacillus licheniformis MTCC 1483 was used to liquefy and saccharify the samples. Response surface modeling was used to optimize the liquefaction step. The technique proposed in this study produces 0.129 g ml1 or 0.32 g per gram of biomass. The cost of producing bioethanol was also calculated, taking into consideration the cost of chemicals, instruments, and running costs. The overall cost of bioethanol synthesized in this study was calculated to be 0.143 dollars per liter of ethanol produced. When compared to the marketing price of ethanol, the price of ethanol produced in the current study was 8.32 times lower [58].

A TEA was undertaken to determine the viability of producing bioethanol from forestry waste with various bark concentrations. Aspen Plus was used to simulate a projected cellulosic ethanol biorefinery in Sweden. The factory was supposed to transform sawdust and shavings, early thinning, tops, fuel logs, hog fuel, branches, and pulpwood into pellets, ethanol, biogas, and electricity. While the overall energy efficiency was between 67 and 69% regardless of the raw material utilized, the cost of producing ethanol varied greatly, with the lowest ethanol selling price ranging from 0.77 to 1.52 USD/L. Except for sawdust and shavings, all forestry residues have a negative net existing value at current market prices under the basic assumptions. The profitability of the raw material reduced as the bark content increased. Sensitivity tests revealed that, at present market pricing, using bark-containing forestry leftovers will not save money when compared to pulpwood until the conversion of lignocellulosic biomass to monomeric sugars improves [59].

References

1. Falkner, R. (2016). The Paris Agreement and the new logic of international climate politics. *International Affairs, 92*(5), 1107–1125.
2. Ritchie, H., & Roser, M. (2020). CO_2 and greenhouse gas emissions. *Our world in data.*
3. Duque, A., Álvarez, C., Doménech, P., Manzanares, P., & Moreno, A. D. (2021). Advanced bioethanol production: From novel raw materials to integrated biorefineries. *Processes, 9*(2), 206.
4. Fulton, L. M., Lynd, L. R., Körner, A., Greene, N., & Tonachel, L. R. (2015). The need for biofuels as part of a low carbon energy future. *Biofuels, Bioproducts and Biorefining, 9*(5),

476–483.
5. Zhang, J., Fisher, T. S., Ramachandran, P. V., Gore, J. P., & Mudawar, I. (2005). *A review of heat transfer issues in hydrogen storage technologies.*
6. Felderhoff, M., Weidenthaler, C., von Helmolt, R., & Eberle, U. (2007). Hydrogen storage: The remaining scientific and technological challenges. *Physical Chemistry Chemical Physics, 9*(21), 2643–2653.
7. Den Boer, E., Aarnink, S., Kleiner, F., & Pagenkopf, J. (2013). Zero emissions trucks. *An overview of state-of-the-art technologies and their potential.*
8. Metz, B., Davidson, O., De Coninck, H. C., Loos, M., & Meyer, L. (2005). *IPCC special report on carbon dioxide capture and storage.* Cambridge University Press.
9. Kim, J. H., Lee, J. C., & Pak, D. (2011). Feasibility of producing ethanol from food waste. *Waste Management, 31*(9–10), 2121–2125.
10. Zaky, A. S., Carter, C. E., Meng, F., & French, C. E. (2021). A preliminary life cycle analysis of bioethanol production using seawater in a coastal biorefinery setting. *Processes, 9*(8), 1399.
11. Neupane, B., Halog, A., & Dhungel, S. (2011). Attributional life cycle assessment of woodchips for bioethanol production. *Journal of Cleaner Production, 19*(6–7), 733–741.
12. Amidon, T. E., Wood, C. D., Shupe, A. M., Wang, Y., Graves, M., & Liu, S. (2008). Biorefinery: Conversion of woody biomass to chemicals, energy and materials. *Journal of Biobased Materials and Bioenergy, 2*(2), 100–120.
13. Amidon, T. E., Bujanovic, B., Liu, S., & Howard, J. R. (2011). Commercializing biorefinery technology: A case for the multi-product pathway to a viable biorefinery. *Forests, 2*(4), 929–947.
14. Amidon, T. E., & Liu, S. (2009). Water-based woody biorefinery. *Biotechnology Advances, 27*(5), 542–550.
15. Therasme, O., Volk, T. A., Cabrera, A. M., Eisenbies, M. H., & Amidon, T. E. (2018). Hot water extraction improves the characteristics of willow and sugar maple biomass with different amount of bark. *Frontiers in Energy Research*, 93.
16. Wood, C. D., Amidon, T. E., Volk, T. A., & Emerson, R. M. (2020). Hot water extraction: Short rotation willow, mixed hardwoods, and process considerations. *Energies, 13*(8), 2071.
17. Runge, T., Wipperfurth, P., & Zhang, C. (2013). Improving biomass combustion quality using a liquid hot water treatment. *Biofuels, 4*(1), 73–83.
18. Volk, T. A., Berguson, B., Daly, C., Halbleib, M. D., Miller, R., Rials, T. G., Abrahamson, L. P., Buchman, D., Buford, M., Cunningham, M. W., & Eisenbies, M. (2018). Poplar and shrub willow energy crops in the United States: Field trial results from the multiyear regional feedstock partnership and yield potential maps based on the PRISM-ELM model. *Gcb Bioenergy, 10*(10), 735–751.
19. Zalesny, R. S., Jr., Berndes, G., Dimitriou, I., Fritsche, U., Miller, C., Eisenbies, M., Ghezehei, S., Hazel, D., Headlee, W. L., Mola-Yudego, B., & Negri, M. C. (2019). Positive water linkages of producing short rotation poplars and willows for bioenergy and phytotechnologies. *Wiley Interdisciplinary Reviews: Energy and Environment, 8*(5), e345.
20. González-García, S., Iribarren, D., Susmozas, A., Dufour, J., & Murphy, R. J. (2012). Life cycle assessment of two alternative bioenergy systems involving Salix spp. biomass: Bioethanol production and power generation. *Applied Energy, 95*, 111–122.
21. González-García, S., Mola-Yudego, B., & Murphy, R. J. (2013). Life cycle assessment of potential energy uses for short rotation willow biomass in Sweden. *The International Journal of Life Cycle Assessment, 18*(4), 783–795.
22. Budsberg, E., Rastogi, M., Puettmann, M. E., Caputo, J., Balogh, S., Volk, T. A., Gustafson, R., & Johnson, L. (2012). Life-cycle assessment for the production of bioethanol from willow biomass crops via biochemical conversion. *Forest Products Journal, 62*(4), 305–313.
23. Stephenson, A. L., Dupree, P., Scott, S. A., & Dennis, J. S. (2010). The environmental and economic sustainability of potential bioethanol from willow in the UK. *Bioresource technology, 101*(24), 9612–9623.
24. Gilani, B., & Stuart, P. R. (2015). Life cycle assessment of an integrated forest biorefinery: Hot water extraction process case study. *Biofuels, Bioproducts and Biorefining, 9*(6), 677–695.

25. Nayak, A., & Bhushan, B. (2019). An overview of the recent trends on the waste valorization techniques for food wastes. *Journal of environmental management, 233*, 352–370.
26. Papadaskalopoulou, C., Sotiropoulos, A., Novacovic, J., Barabouti, E., Mai, S., Malamis, D., Kekos, D., & Loizidou, M. (2019). Comparative life cycle assessment of a waste to ethanol biorefinery system versus conventional waste management methods. *Resources, Conservation and Recycling, 149*, 130–139.
27. Meng, F., & McKechnie, J. (2019). Challenges in quantifying greenhouse gas impacts of waste-based biofuels in EU and US biofuel policies: Case study of butanol and ethanol production from municipal solid waste. *Environmental Science & Technology, 53*(20), 12141–12149.
28. González-García, S., Morales, P. C., & Gullón, B. (2018). Estimating the environmental impacts of a brewery waste–based biorefinery: Bio-ethanol and xylooligosaccharides joint production case study. *Industrial Crops and Products, 123*, 331–340.
29. Guerrero, A. B., & Muñoz, E. (2018). Life cycle assessment of second generation ethanol derived from banana agricultural waste: Environmental impacts and energy balance. *Journal of Cleaner Production, 174*, 710–717.
30. Gnansounou, E., Vaskan, P., & Pachón, E. R. (2015). Comparative techno-economic assessment and LCA of selected integrated sugarcane-based biorefineries. *Bioresource Technology, 196*, 364–375.
31. Bozorgirad, M. A., Zhang, H., Haapala, K. R., & Murthy, G. S. (2013). Environmental impact and cost assessment of incineration and ethanol production as municipal solid waste management strategies. *The International Journal of Life Cycle Assessment, 18*(8), 1502–1512.
32. Ebner, J., Babbitt, C., Winer, M., Hilton, B., & Williamson, A. (2014). Life cycle greenhouse gas (GHG) impacts of a novel process for converting food waste to ethanol and co-products. *Applied Energy, 130*, 86–93.
33. Velásquez-Arredondo, H. I., & Ruiz-Colorado, A. A. (2010). Ethanol production process from banana fruit and its lignocellulosic residues: energy analysis. *Energy, 35*(7), 3081–3087.
34. Prasoulas, G., Gentikis, A., Konti, A., Kalantzi, S., Kekos, D., & Mamma, D. (2020). Bioethanol production from food waste applying the multienzyme system produced on-site by Fusarium oxysporum F3 and mixed microbial cultures. *Fermentation, 6*(2), 39.
35. Matsakas, L., & Christakopoulos, P. (2015). Ethanol production from enzymatically treated dried food waste using enzymes produced on-site. *Sustainability, 7*(2), 1446–1458.
36. Mohd Yusof, S. J. H., Roslan, A. M., Ibrahim, K. N., Syed Abdullah, S. S., Zakaria, M. R., Hassan, M. A., & Shirai, Y. (2019). Life cycle assessment for bioethanol production from oil palm frond juice in an oil palm based biorefinery. *Sustainability, 11*(24), 6928.
37. Nielsen, P. H., Oxenbøll, K. M., & Wenzel, H. (2007). Cradle-to-gate environmental assessment of enzyme products produced industrially in Denmark by Novozymes A/S. *The International Journal of Life Cycle Assessment, 12*(6), 432–438.
38. Gilpin, G. S., & Andrae, A. S. (2017). Comparative attributional life cycle assessment of European cellulase enzyme production for use in second-generation lignocellulosic bioethanol production. *The International Journal of Life Cycle Assessment, 22*(7), 1034–1053.
39. Dunn, J. B., Mueller, S., Wang, M., & Han, J. (2012). Energy consumption and greenhouse gas emissions from enzyme and yeast manufacture for corn and cellulosic ethanol production. *Biotechnology Letters, 34*(12), 2259–2263.
40. Knauf, M., & Kraus, K. (2006). Specific yeasts developed for modern ethanol production. *Sugar Industry/Zuckerindustrie, 131*(11), 753–758.
41. Matsakas, L., Kekos, D., Loizidou, M., & Christakopoulos, P. (2014). Utilization of household food waste for the production of ethanol at high dry material content. *Biotechnology for Biofuels, 7*(1), 1–9.
42. Kiran, E. U., & Liu, Y. (2015). Bioethanol production from mixed food waste by an effective enzymatic pretreatment. *Fuel, 159*, 463–469.
43. Padella, M., O'Connell, A., & Prussi, M. (2019). What is still limiting the deployment of cellulosic ethanol? Analysis of the current status of the sector. *Applied Sciences, 9*(21), 4523.
44. Aditiya, H. B., Mahlia, T. M. I., Chong, W. T., Nur, H., & Sebayang, A. H. (2016). Second generation bioethanol production: A critical review. *Renewable and Sustainable Energy Reviews, 66*, 631–653.

45. Koppram, R., Tomás-Pejó, E., Xiros, C., & Olsson, L. (2014). Lignocellulosic ethanol production at high-gravity: challenges and perspectives. *Trends in biotechnology, 32*(1), 46–53.
46. Galbe, M., Sassner, P., Wingren, A., & Zacchi, G. (2007). Process engineering economics of bioethanol production. *Biofuels*, 303–327.
47. Modenbach, A. A., & Nokes, S. E. (2013). Enzymatic hydrolysis of biomass at high-solids loadings–a review. *Biomass and Bioenergy, 56*, 526–544.
48. Hoyer, K., Galbe, M., & Zacchi, G. (2013). The effect of prehydrolysis and improved mixing on high-solids batch simultaneous saccharification and fermentation of spruce to ethanol. *Process Biochemistry, 48*(2), 289–293.
49. Brown, T. R., Wright, M. M., Román-Leshkov, Y., & Brown, R. C. (2014). Techno-economic assessment (TEA) of advanced biochemical and thermochemical biorefineries. In *Advances in biorefineries* (pp. 34–66). Woodhead Publishing.
50. Roy, P. (2014). *Life cycle assessment of ethanol produced from lignocellulosic biomass: Techno-economic and environmental evaluation* (Doctoral dissertation, University of Guelph).
51. Galbe, M., & Zacchi, G. (2007). Pretreatment of lignocellulosic materials for efficient bioethanol production. *Biofuels*, 41–65.
52. Dao, C. N., Mupondwa, E., Tabil, L., Li, X., & Castellanos, E. C. (2018). *A review on techno-economic analysis and life-cycle assessment of second generation bioethanol production via biochemical processes.*
53. Conde-Mejia, C., Jimenez-Gutierrez, A., & El-Halwagi, M. (2012). A comparison of pretreatment methods for bioethanol production from lignocellulosic materials. *Process Safety and Environmental Protection, 90*(3), 189–202.
54. Petersen, A. M., Van der Westhuizen, W. A., Mandegari, M. A., & Görgens, J. F. (2018). Economic analysis of bioethanol and electricity production from sugarcane in South Africa. *Biofuels, Bioproducts and Biorefining, 12*(2), 224–238.
55. Mupondwa, E., Li, X., & Tabil, L. (2017). Large-scale commercial production of cellulosic ethanol from agricultural residues: A case study of wheat straw in the Canadian Prairies. *Biofuels, Bioproducts and Biorefining, 11*(6), 955–970.
56. Vaskan, P., Pachón, E. R., & Gnansounou, E. (2018). Techno-economic and life-cycle assessments of biorefineries based on palm empty fruit bunches in Brazil. *Journal of Cleaner Production, 172*, 3655–3668.
57. Dempfle, D., Kröcher, O., & Studer, M. H. P. (2021). Techno-economic assessment of bioethanol production from lignocellulose by consortium-based consolidated bioprocessing at industrial scale. *New Biotechnology, 65*, 53–60.
58. Sondhi, S., Kaur, P. S., & Kaur, M. (2020). Techno-economic analysis of bioethanol production from microwave pretreated kitchen waste. *SN Applied Sciences, 2*(9), 1–13.
59. Frankó, B., Galbe, M., & Wallberg, O. (2016). Bioethanol production from forestry residues: A comparative techno-economic analysis. *Applied energy, 184*, 727–736.

Chapter 15
Opportunities and Challenges for Use of Bioethanol as Transport Fuel: A Global Perspective

Abstract Defining the problem of fossil fuel depletion is a global one because it is the major mode of transportation. Bioethanol, which is ethanol made from a variety of bio-based sources, has recently been getting a lot of attention because it can help cut back on carbon dioxide emissions while also reducing the world's growing dependence on fossil fuels. This research examines the effects of ethanol-blend gasoline on engine performance and emission characteristics as well as the use of ethanol as an S.I. engine fuel. In this chapter, the effects of various ethanol blends on engine performance are discussed and reviewed, and showed bioethanol as a fuel in SI engines.

Keywords Bioethanol · Transportation fuel · Effect on engine performance

15.1 Introduction

Several factors make bio-based ethanol a potential biofuel, including the availability of abundant feedstocks, well-established technology, and qualities that are similar to those found in gasoline. Ethanol (C_2H_5OH) is simple liquid alcohol produced from starch-rich grains or lignocellulosic feedstocks or formed by the fermentation of glucose in its natural form. It is estimated that global bioethanol production will drop from 110 billion liters in 2019 to 986 billion liters in 2020 as a result of the pandemic [1, 2]. Ethanol can lower carbon dioxide emissions by 90% and sulfur dioxide emissions by 60–80% when blended with 95% gasoline [3]. According to the EIA (2019), worldwide energy consumption will double in the next three decades, with Asia accounting for the majority of this expansion due to rapid economic growth [4]. In Brazil, almost all bioethanol is produced from sugarcane, which cuts GHG emissions by 78% in comparison to gasoline. Cellulosic biomasses such as bagasse are much more efficient in producing second-generation ethanol (E2G), which reduces GHG, by 86% compared to gasoline [5]. Ethanol use as an alternative fuel has received a lot of attention recently, to enhance the quality of commercial gasoline [6, 7]. Recent studies have indicated that gasoline-ethanol (GE) mixes have been used to boost the research octane number (RON) in recent years [8, 9].

N. K. Aggarwal et al., *Bioethanol Production*, Green Chemistry and Sustainable Technology, https://doi.org/10.1007/978-3-031-05091-6_15

15.2 Transportation Fuel

Domestic bioethanol production has long been promoted in countries such as the US and Brazil. In the US, ethanol/gasoline blends were first pushed as an octane enhancer to replace lead as an environmentally friendly technique. Ethanol is also used in clean-burning gasoline as an oxygenate to lower automotive exhaust emissions. Following Brazil lead in becoming a self-sustaining renewable energy economy, India implemented an ethanol blending scheme and its first biofuel policy in 2003. The strategy planned to introduce 5% ethanol blending in gasoline as a pilot program, to increase the blending ratio to 20% by 2017 [10]. The amount of CO_2-equivalent greenhouse gas emissions from transportation decreased by 43.5 million metric tons in 2016, according to the Renewable Fuel Association (RFA). This is the same as taking 9.3 million automobiles off the road for the entire year (RFA 2017) [11]. The fermentation of sugar and starch crops produces more than 95% of all bioethanol. Corn is the primary source of bioethanol in the United States, while sugarcane is the primary source of bioethanol in Brazil. However, bioethanol manufacturing is dominated by the US and Brazil, whereas biodiesel is primarily produced in the European Union. More than 30 countries have established or are considering establishing gasoline-ethanol programs [12]. In spark-ignition (SI) engines, ethanol is normally blended with pure gasoline at a predetermined percentage and fed into the inlet manifold or directly into the engine cylinders [13]. Several countries' efforts to reduce their reliance on crude oil reserves, develop agricultural-based economies, and boost air quality have resulted in consistent rises in global ethanol output over the previous year [14]. Ethanol is colorless, neutral, transparent, volatile, and easily mixable with non-polar solvents, with a 34% wt O_2 level [15]. This ethanol is used in 10% gasoline blends known as gasohol, a name that is gradually being phased out in favor of ethanol/gasoline blends or E10. The E10 blend and lesser ethanol percentage combinations have been used in a number of nations, and their adoption has been sparked by various global energy crises since the 1973 oil crisis [16]. In addition, E10 and other ethanol blends are expected to be effective in reducing the US reliance on foreign oil and, in the correct circumstances, can reduce carbon monoxide emissions by 20–30% (Fig. 15.1).

The EPA approved a waiver in October 2010 allowing up to 15% ethanol combined with gasoline to be marketed just for cars and light pickup trucks with a model year of 2007 or later, which accounts for around 15% of vehicles on US roads. The Renewable Fuels Association estimates that the theoretical blend wall for ethanol usage in the United States would be 17.5 billion gallons annually if all 2001 or newer cars and trucks were to utilize E15, which now covers 62% of vehicles on the road. The EPA launched a regulation in October 2018 to authorize year-round manufacturing of E15. E20 has 20% ethanol, whereas E25 has 25% ethanol. In certain cases, smaller amounts of ethanol may operate, but with a few exceptions, automobiles traveling to adjacent South American nations have shown that they cannot run smoothly on pure gasoline because it produces engine banging. In mid-2003, vehicles that could run on any combination of gasoline E20-E25 up to 100%

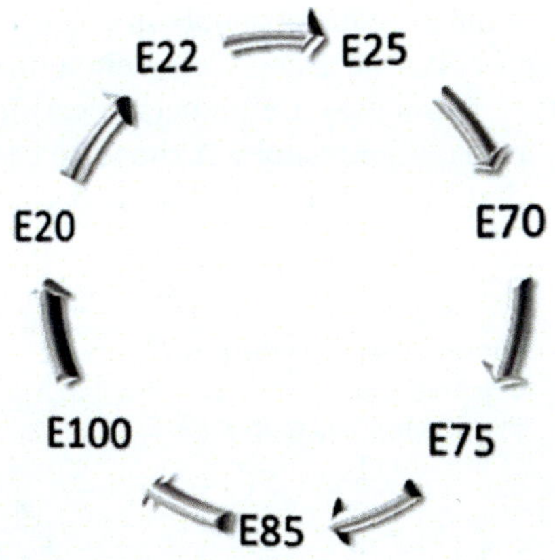

Fig. 15.1 Variety of blends available in Brazil

hydrous ethanol became available. Winter mixes for E85 flexible-fuel vehicles in the United States and Sweden include E70 and E75, which are still available at the pump as E85 [17]. E85, the maximum ethanol concentration of any fuel available in the United States or Europe, is commonly used in flexible-fuel vehicles. Countries like Australia (up to 10%), Brazil (25%), Canada (up to 10%), Sweden (5%), and the United States all use varying percentages of ethanol in their gasoline blends (up to 10%). However, automobile manufacturers in the United States have agreed that the use of gasoline containing up to 10% ethanol will not impact the warranties of their vehicles [18, 19]. Newer flexible-fuel vehicles in Brazil can run on either pure hydrous ethanol (E100) or a blend of E20 to E25 gasoline (a mixture manufactured with anhydrous ethanol), which is the national mandated blend. Ethanol reduced evaporative pressure (in comparison to gasoline) presents issues when cold starting the engine at temperatures below 15 °C., which limits the usual operation of an E100 vehicle. India has already implemented a strategy of ethanol mixes in gasoline to minimize pollution and import costs [20]. As a result, promoting ethanol as a vehicle fuel presents a significant opportunity. In India, the primary source of ethanol is sugarcane through molasses, a byproduct of the sugarcane juice to sugar conversion process. Molasses, a byproduct of the sugarcane juice to sugar conversion process, is the primary source of ethanol in India. The government of India set up a national strategy on biofuels in 2009, and it has a goal of selling petrol that has at least 5% ethanol in it. This goal may be increased in the future [20, 21]. From April 2020 to January 2021, the blending rate in Tamil Nadu was 1.91 percent per liter. In January 2021, it reached 6.73%, compared to a target of 10% per liter. Currently, a liter of ethanol costs 61₹. However, the price is determined by internal fuel prices. However, ethanol blending has had little effect on the price of gasoline, sometimes known as "motor spirit." The customer just pays for a liter of petrol [22]. The Indian government said today that the country's just-completed 2020–21 (Dec–Nov) supply year had the highest-ever ethanol blend in India at 8.1%. The distilleries supplied 3.03 billion liters of ethanol to the oil marketing companies in the ethanol supply year 2020–21 (Dec–Nov), which is the biggest amount ever. According to the food ministry, ethanol supply in 2013–14 was just 0.38 billion liters, with a mixing level of only 1.53%, increasing eight times in the past seven years [23] (Fig. 15.2).

Ethanol as a high-octane gasoline additive was considered by Rasskazchikova et al. [24]. Ethanol was shown to be the most potential octane-enhancing component, despite its high cost. This is because ethanol-containing fuel is less toxic, has less

Fig. 15.2 **a** Physicochemical properties of gasoline, **b** Physicochemical properties of bioethanol

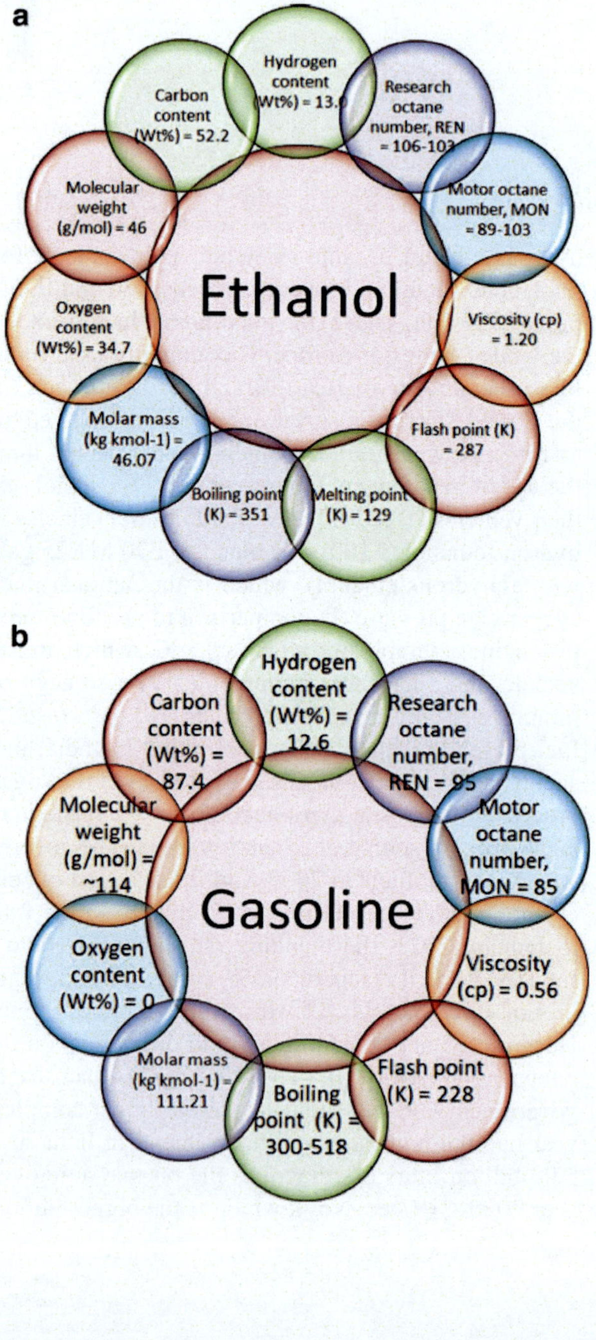

impact on the environment, and comes from renewable sources. Studying the anti-knock properties and Reid vapor pressure (RVP) of gasoline was done by Silva et al. [25] using ethanol and other additives. Increased RVP and octane ratings were achieved by adding up to 25% ethanol by volume to gasoline. It is possible to use ethanol as a fuel for automobiles, either as a stand-alone fuel or in combination with gasoline. The compatibility of a SI engine with ethanol-blended gasoline containing 10% ethanol was investigated by Shanmugam et al. [26]. There was a 13% drop in diluted carbon monoxide emissions because of the blend. There was also a 19% drop in hydrocarbon emissions, a 2% drop in carbon dioxide emissions, and a 16% rise in nitrogen oxide emissions. Chen et al. [27] investigated the impact of ethanol-gasoline blended fuel on SI engine cold-start emissions. The engine could be started reliably with E5, E10, E20, and E30, according to the results. With more ethanol in the blends than 20% ethanol, hydrocarbon and carbon monoxide emissions are reduced dramatically. The E40's engine idle became unsteady. As a result, it was found that the ethanol percentage in gasoline should be at least 20% but no more than 30% for the best cold-start emissions. Costa and Sodré [28] examined the impact of compression ratio on the efficiency of a SI engine running on a mixture of 78% gasoline and 22% ethanol or hydrous ethanol (E100). According to the findings, higher compression ratios enhanced engine efficiency for the fuel blends over the whole speed range, with significant effects when hydrous ethanol was used.

Jia et al. [29] used chassis dynamometers to investigate the emission characteristics of a motorcycle engine in various driving modes. The results showed that using E10 reduced carbon monoxide and hydrocarbon emissions in the engine exhaust when compared to using unleaded gasoline, but that the outcome of ethanol on nitrogen oxide emissions was not significant. Ethylene, acetaldehyde, and ethanol emissions were also higher in E10-fueled motorcycle engines than in gasoline engines that did not use E10. Schifter et al. [30] examined the impact of mid-level gasoline-ethanol blends (0–20 vol percent ethanol) on engine performance and exhaust emissions in a single-cylinder engine, concluding that mixes up to 10 vol percent had marginal effects on combustion rates when compared to gasoline, but blends up to 20 vol percent slowed the combustion process and enhanced cyclic dispersion. E-85 is safe for the environment [31]. It contains the most oxygen of any fuel currently available, allowing it to burn cleaner than regular gasoline. E85 helps to minimize ozone and CO emissions, as well as air toxins like benzene. When compared to vehicles that use regular unleaded gasoline, E-85 cars function well and emit significantly less pollution. Ethanol is one of only two liquid fuels that combat global warming due to its source of raw materials. Corn transforms carbon dioxide into oxygen as it grows. Automobile manufacturers are developing vehicles that can run on a variety of fuels. These vehicles have a purchase price that is comparable to the base price of gasoline models. Flexible-fuel cars should last a little longer than gasoline cars because E-85 is a cleaner-burning fuel. In an engine, ethanol (E100) consumes approximately 34% more than gasoline (the energy per volume unit is 34% lower). The higher the compression ratio, the more power, and better fuel economy you get from an ethanol-only engine. When properly operated, modern computerized automobiles outperform non-computerized vehicles. The computerized fuel system in the car can adapt to changes in operating conditions or fuel type, resulting in better performance. Deposits on port fuel injectors have been linked to some of the chemicals used in the production of gasoline, such as olefins. Detergent additives in today's gasoline are designed to prevent deposits in fuel injectors and valves.

15.3 Effect on Engine Performance

Because of the use of blends, the gasoline engine's efficiency has grown significantly. The thermal efficiency of the brakes increased with increasing brake power when utilizing E5 and E10 blends at both vehicle speeds. The decrease in brake-specific fuel consumption is a major contributor to this condition. At speeds of 100 km per hour, the thermal efficiency increases of E5, E10 are 1.9% and 2.5%, respectively, when compared to unleaded gasoline. E10's thermal efficiency increases by 0.4% at a vehicle speed of 80 km per hour, except for E5. When compared to unleaded gasoline, the thermal efficiency of E5 is 0.8% lower. Because the E10 blend has a higher oxygen rate than the E5, the combustion improves and the thermal efficiency rises. It has been determined that the use of blends has a favorable impact on the fuel's octane rating and that increased use of blends has boosted the engine's overall braking thermal efficiency. With E5 and E10 blends, the braking thermal efficiency of the engine is higher than with gasoline [32, 33]. When compared to gasoline fuel, the volumetric efficiency of blended fuel is said to be higher. The improved breathing capacity of the engine can be linked to the higher temperature generated in the cylinder when mixes burn in it, resulting in cleaner and higher combustion in the cylinder. The higher volumetric efficiency is due to the impression of increased evaporative cooling brought on by the blends. The volumetric efficiency is estimated to be 5—10% higher than with gasoline [34, 35]. Because there is always some incomplete combustion of the fuel inside the engine, the CO content cannot be reduced to zero. The amount of CO in the exhaust is also controlled by the engine's operating temperature. In many circumstances, the lower temperature actually increases the amount of carbon monoxide in the exhaust. The amount of emissions changes depending on the engine's compression ratio. Carbon monoxide emissions can be lowered by 30% when ethanol is added to gasoline at a 10% concentration. An evidential statement on the lowering of carbon monoxide with the introduction of various mixes was discovered in this inquiry [6, 36]. At low load and low-speed conditions, studies conducted on the same engine have demonstrated a reduction in carbon monoxide of up to 15% with ethanol blends of up to 20%. Up to 10% ethanol blends in gasoline resulted in a modest reduction in carbon monoxide. With ethanol blends, hydrocarbon emissions were reduced by up to 20% at high load and high-speed situations. When ethanol blends were used instead of base gasoline, nitrogen oxide emissions were reduced by 40% when running at part load. The equivalency ratio, peak combustion temperature, and engine operating conditions all influence the generation of nitrogen oxide emissions. The high latent heat of vaporization, low combustion peak temperature, and lower equivalence ratio all help to reduce nitrogen oxide emissions when the machine is running at a lower load [37]. At high loads, ethanol-gasoline blends had longer burn duration than straight gasoline during wide-open throttle operation of a passenger automobile engine. This could be owing to the engine's increased speed and the high latent heat of vaporization. Furthermore, leftover gases in the cylinder restrict the flame speed at high speeds. When ethanol blends were compared to pure gasoline at the same component loads, there was a slight rise in combustion time of up to 20%. At part loads, there was no change in the indicated mean effective pressure or peak pressure. Compared to base gasoline, ethanol blends increased indicated mean effective pressure by 2.2% and peak pressure by 3.5% when run at wide-open throttle. Fuel had no effect on the peak heat release rate under part loads, but E20 had a 7% higher rate than base gasoline under wide-open throttle [37, 38].

15.4 Summary

The study's goal is to gather data on national and worldwide results and experiences on the usage of ethanol-in-gasoline blends as fuels in spark-ignition engines. Blending ethanol with gasoline to cut down on fossil carbon dioxide emissions (and thus the greenhouse effect) from cars is the main reason to use bioethanol. A blend of biofuels and a petroleum-based fuel has both advantages and disadvantages. Even small percentage increases in biofuel use can result in a lot of total gasoline substitution, and the existing fuel distribution infrastructure can be used almost unchanged. The study findings suggest that ethanol can be used as a fuel and blended with gasoline in a manner that is well-established. According to the findings of this study, adding ethanol to gasoline can also increase engine performance and reduce carbon monoxide and hydrocarbon emissions. However, it has the potential to raise emissions of carbon dioxide and nitrogen oxides.

References

1. The Future of Biofuels: A Global Perspective. Available online: https://www.ers.usda.gov/amber-waves/2007/november/thefuture-of-biofuels-a-global-perspective/. Accessed on 19 Septe 2021.
2. Global Ethanol Production by Country or Region. Available online: https://afdc.energy.gov/data/10331/. Accessed on 19 Septe 2021.
3. Halder, P., Azad, K., Shah, S., & Sarker, E., (2019). Prospects and technological advancement of cellulosic bioethanol ecofuel production. In *Advances in eco-fuels for a sustainable environment* (pp. 211–236). Woodhead Publishing.
4. EIA. (2019). Today in Energy. Accessed 30 June 2020. https://www.eia.gov/todayinenergy/detail.php?id=41433
5. Delgado, F., Roitman, T., de Sousa, M. E. (2017). Biofuels. *Cadernos FGV energia* 4:(8).
6. Koç, M., Sekmen, Y., Topgül, T., & Yücesu, H. S. (2009). The effects of ethanol–unleaded gasoline blends on engine performance and exhaust emissions in a spark-ignition engine. *Renewable Energy, 34*(10), 2101–2106.

7. Amine, M., & Barakat, Y. (2019). Properties of gasoline-ethanol-methanol ternary fuel blend compared with ethanol-gasoline and methanol-gasoline fuel blends. *Egyptian Journal of Petroleum, 28*(4), 371–376.
8. Wibowo, C.S., Sugiarto, B., Zikra, A., Budi, A., & Mulya, T. (2018). The Effect of Gasoline-Bioethanol Blends to The Value of Fuel's Octane Number. In *E3S Web of Conferences* (Vol. 67, p. 02033). EDP Sciences.
9. Adian, F., Sugiarto, B., Wibowo, C.S., Primayandi, D.D., Hargiyanto, R., & Krisnanto, H. (2020, May). Comparison of the effect of 10% ethanol addition in 88 and 98 gasoline RON on motorcycle engine performance. In *AIP Conference Proceedings* (Vol. 2230, No. 1, p. 050003). AIP Publishing LLC.
10. Das, S., 2018. Achievements and misses of the Indian national policy on biofuels 2009. Achievements and misses of the Indian national policy on biofuels 5–30.
11. RFA-Renewable Fuels Association. (2019). Ethanol industry outlook. https://ethanolrfa.org/wp-content/uploads/2019/02/RFA2019Outlook.pdf.
12. Rosillo-Calle, F., & Walter, A. (2006). Global market for bioethanol: Historical trends and future prospects. *Energy for Sustainable Development, 10*(1), 20–32.
13. Bae, C., & Kim, J. (2017). Alternative fuels for internal combustion engines. *Proceedings of the Combustion Institute, 36*(3), 3389–3413.
14. Mahmudul, H. M., Hagos, F. Y., Mamat, R., Adam, A. A., Ishak, W. F. W., & Alenezi, R. (2017). Production, characterization and performance of biodiesel as an alternative fuel in diesel engines–A review. *Renewable and Sustainable Energy Reviews, 72*, 497–509.
15. Masum, B. M., Masjuki, H. H., Kalam, M. A., Fattah, I. R., Palash, S. M., & Abedin, M. J. (2013). Effect of ethanol–gasoline blend on NOx emission in SI engine. *Renewable and Sustainable Energy Reviews, 24*, 209–222.
16. EAIP (Earth Policy Institute). (2001). Setting the ethanol limit in petrol. Environment Australia issues paper.
17. Bajpai, P. (2020). *Developments in bioethanol*. Springer Nature.
18. Kheiralla, A. F., Tola, E., & Bakhit, J. M. (2017). Performance of ethanol-gasoline blends of up to E35 as alternative automotive fuels. *Advances in Bioresearch, 8*(5).
19. Egebäck, K., Henke, M., Rehnlund, B., Wallin, M., &Westerholm, R. (2005). Blending of ethanol in gasoline for spark ignition engines problem inventory and evaporative measurements. MTC AB Report No. MTC 5407, ISSN: 1103–0240, ISRN: ASB-MTC-R—05/2—SE, p 133.
20. GoI (2009). National policy on biofuels. Ministry of New and Renewable Energy, Government of India, New Delhi.
21. Pohit, S., Biswas, P. K., Kumar, R., & Jha, J. (2009). International experiences of ethanol as transport fuel: Policy implications for India. *Energy Policy, 37*(11), 4540–4548.
22. Ramakrishnan, D. H. (2021). What is Ethanol-blended petrol? Should we be concerned?. Accessed 28 Feb 2021. https://www.thehindu.com/news/national/tamil-nadu/what-is-ethanol-blended-petrol-should-we-be-concerned/article33954659.ece.
23. Mukherjee, S. (2021). India achieves highest-ever ethanol blend of 8.1% with petrol in 2020–21. Accessed 8 Dec 2021. https://www.business-standard.com/article/economy-policy/india-achieves-highest-ever-ethanol-blend-of-8-1-in-2020-21-121120700966_1.html.
24. Rasskazchikova, T. V., Kapustin, V. M., &Karpov, S. A. (2004). Ethanol as high–octane additive to automotive gasolines. production and use in russia and abroad. *Chemistry and Technology of Fuels and Oils, 40*(4), pp.203–210.
25. Da Silva, R., Cataluna, R., de Menezes, E. W., Samios, D., & Piatnicki, C. M. S. (2005). Effect of additives on the antiknock properties and Reid vapor pressure of gasoline. *Fuel, 84*(7–8), 951–959.
26. Shanmugam, R. M., Saravanan, N., Srinivasan, L., Hosur, V., & Sridhar, S. (2009). *An Experimental investigation on 1.4 L MPFI gasoline engine to study its performance, emission and compatibility with E10 fuel* (No. 2009-01-0611). SAE Technical Paper.
27. Chen, R. H., Chiang, L. B., Chen, C. N., & Lin, T. H. (2011). Cold-start emissions of an SI engine using ethanol–gasoline blended fuel. *Applied Thermal Engineering, 31*(8–9), 1463–1467.

28. Costa, R. C., & Sodré, J. R. (2011). Compression ratio effects on an ethanol/gasoline fuelled engine performance. *Applied Thermal Engineering, 31*(2–3), 278–283.
29. Jia, L. W., Shen, M. Q., Wang, J., & Lin, M. Q. (2005). Influence of ethanol–gasoline blended fuel on emission characteristics from a four-stroke motorcycle engine. *Journal of Hazardous Materials, 123*(1–3), 29–34.
30. Schifter, I., Diaz, L., Rodriguez, R., Gómez, J. P., & Gonzalez, U. (2011). Combustion and emissions behavior for ethanol–gasoline blends in a single cylinder engine. *Fuel, 90*(12), 3586–3592.
31. Niven, R. K. (2005). Ethanol in gasoline: Environmental impacts and sustainability review article. *Renewable and sustainable energy reviews, 9*(6), 535–555.
32. Sharma, T. K. (2015). Performance and emission characteristics of the thermal barrier coated SI engine by adding argon inert gas to intake mixture. *Journal of advanced research, 6*(6), 819–826.
33. Eyidogan, M., Ozsezen, A. N., Canakci, M., & Turkcan, A. (2010). Impact of alcohol–gasoline fuel blends on the performance and combustion characteristics of an SI engine. *Fuel, 89*(10), 2713–2720.
34. Elfasakhany, A. (2015). Investigations on the effects of ethanol–methanol–gasoline blends in a spark-ignition engine: Performance and emissions analysis. *Engineering Science and Technology, an International Journal, 18*(4), 713–719.
35. Mittal, N., Athony, R. L., Bansal, R., & Kumar, C. R. (2013). Study of performance and emission characteristics of a partially coated LHR SI engine blended with n-butanol and gasoline. *Alexandria Engineering Journal, 52*(3), 285–293.
36. Palmer, F.H. (1986, November). Vehicle performance of gasoline containing oxygenates. In *International conference on petroleum based fuels and automotive applications. IMECHE conference publications 1986–11. Paper no C319/86.*
37. Ramadhas, A. S., Singh, P. K., Sakthivel, P., Mathai, R., & Sehgal, A. K. (2016). *Effect of ethanol-gasoline blends on combustion and emissions of a passenger car engine at part load operations* (No. 2016–28–0152). SAE Technical Paper.
38. Singh, P. K., Ramadhas, A. S., Mathai, R., & Sehgal, A. K. (2016). Investigation on combustion, performance and emissions of automotive engine fueled with ethanol blended gasoline. *SAE International Journal Of fuels and lubricants, 9*(1), 215–223.